KT-522-858
ROOM

THE EVALUATIVE IMAGE OF THE CITY

To my wife
Judy D. Nasar

Jack L. Nasar

THE EVALUATIVE IMAGE OF THE CITY

SAGE Publications
International Educational and Professional Publisher
Thousand Oaks London New Delhi

For information:

SAGE Publications, Inc.
2455 Teller Road
Thousand Oaks, California 91320
E-mail: order@sagepub.com

SAGE Publications Ltd.
6 Bonhill Street
London EC2A 4PU
United Kingdom

SAGE Publications India Pvt. Ltd.
M-32 Market
Greater Kailash I
New Delhi 110 048 India

Printed in the United States of America

Library of Congress Cataloging-in-Publication Data

Nasar, Jack L.
The evaluative image of the city / by Jack L. Nasar.
p. cm.
Includes bibliographical references and index.
ISBN 0-8039-5447-6 (cloth : acid-free paper). — ISBN 0-8039-5448-4 (pbk. : acid-free paper)
1. City planning—Psychological aspects. 2. Urban beautification. 3. Urban ecology. 4. Architecture and society. 5. Architecture—Aesthetics. I. Title.
HT166.N35 1997
307.1'216—dc21 97-21151

98 99 00 01 02 03 10 9 8 7 6 5 4 3 2 1

Acquiring Editor:	Catherine Rossbach
Editorial Assistant:	Kathleen Derby
Production Editor:	Diana E. Axelsen
Production Assistant:	Denise Santoyo
Typesetter/Designer:	Danielle Dillahunt
Cover Designer:	Ravi Balasuriya

Contents

Preface

This book examines city appearance, whether appearance matters, and what we can do to make our cities look better. Cities can evoke a sense of delight and pleasure. Their ambiance arises from social and cultural factors. The apparent "politeness" of the British, the "rudeness" of the New Yorker, or changes in atmosphere from immigration in cities such as London, Sydney, or Miami reflect sociocultural influences on the ambiance. The ambiance of a city also arises from its physical form. Good city appearance is not an abstract aesthetic phenomenon; it depends on the evaluations of the people who regularly experience the city. If they appreciate and value the appearance, the city has a good visual form.

City form evolves from many actions by many entities, both public and private. Through regulations, design review, and individual development decisions, we can shape the visual form of our communities for better or for worse. We do not do so just for the sake of visual form. We do so to improve the community's meaning and appearance for the many people who experience it.

In considering community appearance, this book looks at resident and visitor evaluations of the visual form of two American cities: Knoxville and Chattanooga, Tennessee. It also describes evaluations of other places in the United States and Canada. It suggests methods that allow citizens to improve the visual character of their communities, and it offers guidelines for design. The book introduces the concept of the evaluative image, the rationale for studying it, and the need for a scientific approach.

It also presents case studies, design principles derived from the research, and discussion of the application of the findings and methods for city design.

Many groups have an interest in shaping community form: citizens, members of design review or planning commissions, members of chambers of commerce, students, and professionals in such fields as architecture, landscape architecture, environmental design, city planning, environmental psychology, real estate, public policy, and urban design. I hope this book can help such groups make our cities more enjoyable places to live.

Acknowledgments

Portions of this book have appeared in different form elsewhere (Nasar, 1990, 1994). Readers may also recognize the influences of scholars from various fields—planning, urban design, psychology, and anthropology—on this work. The book builds on the influential ideas of the planner and urban designer Kevin Lynch (1960) and his seminal book *The Image of the City.* It links his ideas on mental maps to psychological theory, methods, and findings of Daniel Berlyne (1971), Rachel and Stephen Kaplan (1989), Amos Rapoport (1990b), Roger Ulrich (1983), and Joachim F. Wohlwill (1976). I thank my father, who taught me the value of popular culture; Oscar Newman, who introduced me to the idea of empirical research for urban design; Roger Downs, who introduced me to the concepts of mental maps; and Joachim F. Wohlwill, who served as a model shaping my approach to research and theory.

The Knoxville's EXPO Community Appearance Committee and the University of Tennessee School of Architecture provided support for the research in Knoxville. The Lindhurst Foundation Center for Environmental Action provided financial support for the work in Chattanooga. As teaching assistants, Samuel C. Matthews and James Fentress helped guide the Knoxville and Chattanooga studies. The following students worked on those studies: Brent Aston-Wash, Delores Beets, Behran Bonyadi, James Brewer, J. Mark Buck, Mike Carpenter, Larry Craighead, Randy Dender, John Felton, Joanne Grimes, Connie Hall, Lynn Hampton, Mark Hawks, Mehran Heidari, William Johnson, Hamid Khanli, Rod Lee, Gavin McFarlane, Jeff Rindin, Mehran Saffari, Dane

Swindell, Keith Walters, and Scott Webb. The Ohio State University Office of Research and Graduate Studies provided financial support for the signscape research, as part of a larger project with Steven I. Gordon. Abir Mullick, Hugo Valencia, Al Rezoski, and Adam Prager assisted on the sign project. Richard Makley and Tom Miller helped with the street image project. Junmo Kang and Yen Maw-Chang worked on the German Village Project. Kazunori Hanyu worked on the Columbus neighborhood study and provided useful information on his Tokyo work. Hundreds of people participated as respondents in the various studies. I am grateful to them all.

I am also grateful to Gary Moore, Robert Marans, Kimberly Devlin, and Stephen Kaplan for their comments on portions of this book and to Sidney Bower, Gary Evans, Fred Hurand, Amos Rapoport, and Erv Zube for their comments on early drafts of the complete book, and to the wonderful editors and staff at Sage—Catherine Rossbach, Diana Axelsen, Deirdre MacDonald Greene, Jennifer Morgan, Kathleen Derby, and Danielle Dillahunt—who helped shape the manuscript into its present form. Finally, I am grateful to City and Regional Planning at the Ohio State University for giving me the time to work on this book.

1

The Evaluative Image of the Environment

Think about your experience with your city, town, or neighborhood. Can you think of places that grab your attention and leave you feeling pleasure or delight, places that you experience as fearful or unsafe, or places that you experience as restful? As the architect anthropologist Amos Rapoport (1993) notes, cities and parts of cities have an *ambiance,* a sensory quality or character that we can easily feel. Though your feelings depend on the situation (such as time of day, who is with you, and familiarity), environmental cues that may have escaped your consciousness shape your evaluation, feelings, inferences, and behavior. We respond to what appears before us, to inferences derived from visual cues, and to our recall of places. We see meanings in places. The geographers Peter Gould and Rodney White (1974) refer to this mental landscape of meanings as an *invisible landscape* that shapes our behavior. Visual quality can have powerful effects on our experience and the delight we take in our surroundings. For the city environment, multiply the individual experience by the millions of people who experience cities daily.

The form of cities in the United States results from many acts by many individuals, businesses, developers, and agencies both public and private. This host of actors following cultural rules over time produces a style that Rapoport (1993) calls a "recognizable cultural landscape" (p. 36).

Thus, many parts of U.S. cities—such as the roadside strip, the mall, the suburb—look alike across the country, though certain areas may have distinctive features. Although cultural rules shape our landscape, the aggregate outcome of individual decisions may yield a disagreeable community character for the millions of commuters, shoppers, visitors, and others who experience the city landscape. Garrett Hardin (1968) refers to a *tragedy of the commons,* where what appears good to each individual becomes harmful to the community. Community appearance may suffer from a form of the tragedy of the commons: Alone, each new building, sign, or element may appear desirable or harmless, but the aggregate looks ugly and disturbing. Some U.S. cities such as Boston or San Francisco, or parts of cities, may convey a positive image. Otherwise, dullness and disorder prevail in the hodgepodge of buildings and parking lots downtown, the chaotic signs and traffic in commercial strips, the billboards competing for attention along highways, and the unsightly development intruding in areas of natural beauty (Tunnard & Pushkarev, 1981).

The shaping of city form differs from visual arts such as painting or sculpture. City form continually changes as a result of a multitude of actions, and it affects many ordinary people in their day-to-day activities. It is public. Though we may accept the idea of "high" visual arts that appeal to a narrow audience who choose to visit a museum, city form and appearance must satisfy the broader public who regularly experiences it. To know about the appeal of the city form, one must measure people's responses.

Likability

The city landscape may have value as a source of delight to people and a possible restoration from the stresses of everyday life. Toward this end, the shaping and reshaping of the city "should be guided by a 'visual' plan: a set of recommendations and controls . . . concerned with visual form on the urban scale" (Lynch, 1960, p. 116). To devise such a plan, we need to know how the public evaluates the cityscape and what meanings they see in it: the *evaluative image* of the city.

Just as we weigh objects to find how light or heavy they are, we can measure preferences to determine the degree to which people like or dislike various areas of a city. This book describes research aimed at uncovering this information. It considers the visual quality of the American city by studying the shared public image of the city and its parts. It focuses on one aspect of the evaluative image and visual quality: the likability of the cityscape. *Likability* refers to the probability that an environment will evoke a strong and favorable evaluative response among the groups or the public experiencing it. Inhabitants of a city with a good evaluative image find pleasure in the appearance of its memorable and visible parts. Thus, rather than treating city appearance as an aesthetic object in itself, this book considers city appearance as evaluated by the public who experiences it.

Americans live with visual disorder in the environment. We may have learned to accept it, adapt to it, or turn a blind eye to it, but I believe we would find more enjoyment in more agreeable surroundings. Such surroundings could also convey desirable meanings to visitors. This book argues that we underestimate the importance of a city's evaluative image or likability. I present several studies of urban likability and discuss the use of the findings and methods to improve city appearance. The book represents a preliminary effort. The findings reflect certain populations, places, measures, and conditions. Confirming whether these apply to other situations will require further study. The use of the various methods allows for such testing and the development of situation-specific guidelines, however.

Many factors other than the evaluative image make for a successful city. By stressing the evaluative image, I do not mean to imply that a facelift alone will solve urban problems. The evaluative image or likability has a fundamental value, however. Research (discussed in Chapter 4, "The Elements of Urban Likability") shows widespread agreement on preferred environmental features, and it confirms the centrality of environmental evaluation and appearance to humans. Appearance and meaning are not separate from function but central to it. The disagreeable appearance of our cities goes beyond an absence of emotional satisfaction and abstract notions of good aesthetic form. In their incompatibility with human activity, appearance and meaning may heighten sensory overload, fear, and stress. Disorder in the form of physical incivilities such as dilapidation, graffiti, litter, and abandoned buildings can evoke a sense of anxiety and fear suggesting a threat to survival (Nasar, 1983; Perkins, Meeks, & Taylor, 1992; Skogan & Maxfield, 1981; Taylor, 1989; Taylor, Shumaker, & Gottfredson, 1985; Warr, 1990). Disorder also may affect rates of crime (Perkins, Wandersman, Rich, & Taylor, 1993; Taylor, 1989). With careful attention to improving the evaluative image, we can resolve these problems and enhance well-being.

Building the Evaluative Image

The evaluative image arises from the person and the environment and the ongoing interaction between the two (Figure 1.1). It may vary with biology, personality, sociocultural experience, adaptation levels, goals, expectations, and internal and external factors. The environment has many attributes. Observers, depending on both internal and environmental factors, overlook some attributes, attend to others, and evaluate what they see. This evaluation may involve varying amounts of mental activity. It may involve feelings relating directly to the structure of the form and requiring little to no cognition or mental activity. Figure 1.1 shows this as environmental perception. Environmental evaluation may also arise from the content meaning of the form. This requires mental activity to (1) recognize the content (such as a park); (2) draw inferences about it and place it into a mental framework (a safe park); and (3) evaluate it. Figure 1.1 shows this as environmental cognition. In sum, we filter our evaluative response through the lens of our perception and cognition of the environment.

Because of each human's uniqueness and unique experiences, the evaluative image of a place or city will vary across observers. This assumption leads to little analysis useful for urban design. Science attempts to bring order to experiences that appear varied by finding agreement or universal principles. Think of the possible levels of agreement as varying from absolutely no agreement to perfect agreement. Although we do not share the same evaluative images with one another (perfect agreement), we do have some overlaps in our evaluative images. Shared physical reality, physiology, and culture produce areas of agreement.

Humans communicate verbally, in written and spoken words, and they communicate nonverbally, through such things as clothing, facial expressions, gestures, and relative positions. The built environment represents a channel of nonverbal communication (Rapoport, 1990b), though it has

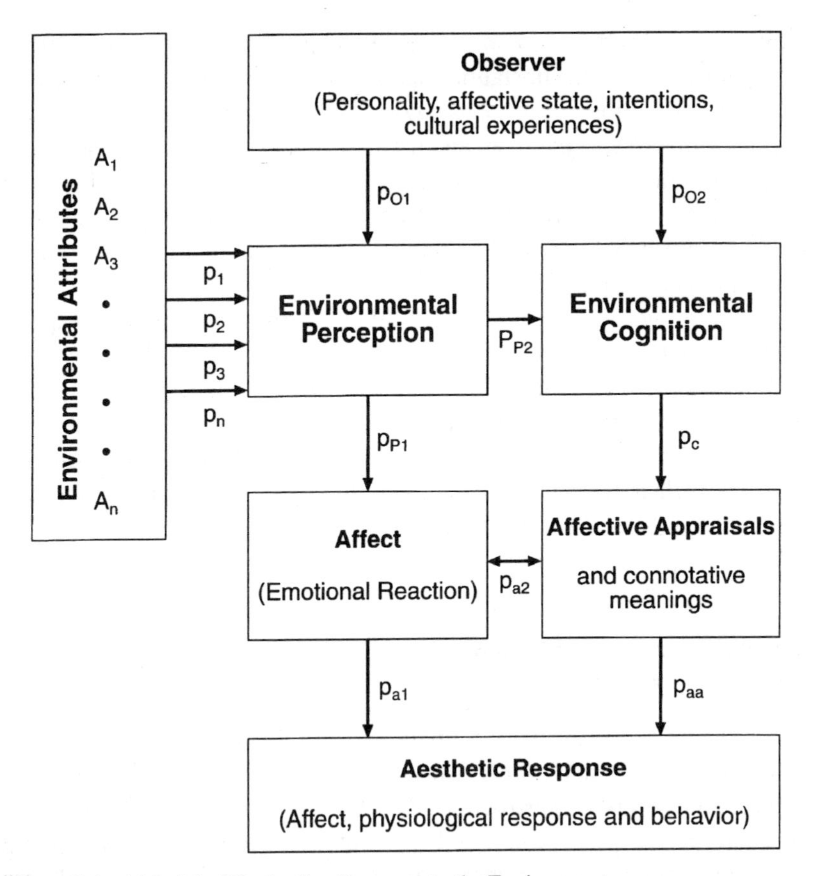

Figure 1.1. A Model of Evaluative Response to the Environment
Graphic by David Miller.

a verbal dimension in the text of traffic signs and outdoor advertising. Individuals experiencing similar sets of messages from the physical environment develop similar value systems toward it. Groups of individuals of similar cultural or socioeconomic characteristics share common meanings and evaluative images (Michelson, 1976). The commonalities may become more diffuse as one moves from a small area to a larger one or from a sociocultural group to a broader population, but the broader population shares evaluative images for the larger areas.

Any given form will vary in the probability of its evoking a favorable evaluative image among various observers. Acknowledging possible individual differences in response, architect planner Kevin Lynch (1960) stresses the areas of substantial agreement. He argues that the group image, representing a consensus among significant numbers of individuals, has value to city planners and urban designers. These professionals, who shape the physical environment for use by many people, need to know the likely effects of the manipulation on the people who experience it. Therefore, I will stress the "public" evaluative responses shared by large numbers of people.

Although various cultures and individuals may have unique experiences with their environment, research confirms that the processes underlying their evaluations and the elements of their evaluations have much in common. Studies point to several visual features as relating to social meanings and preferences, and findings on the evaluative images of cities reflect five of those features: naturalness, upkeep, openness, order, and historical significance.

Identity, Structure, and Likability

The concept of an evaluative image extends Lynch's (1960) work on cognitive maps. Lynch sought community consensus on the elements that enhance the identity and structure of a city—its *imageability* or *legibility.* Imageability helps us orient and find our way around, thus enhancing our enjoyment of a city. Lynch feels that people should have a reliable knowledge about the things present in their environment, and he assumes that people will more likely know, and so use, an environment that is easy to read, or legible. This makes legibility a legitimate purpose of design. The extensive research on cognitive maps, orientation, and wayfinding has yielded useful information for the design of cities.

According to Lynch (1960), the environmental image has three parts: *identity, structure,* and *meaning.* We recognize or identify objects, we see a recognizable pattern, and we draw emotional value in relation to them. Object recognition, which depends on distinction or a noticeable difference, represents identity. Recognizing the pattern of relationships organizing the objects represents the structure. Meaning has three levels: A lower-level meaning, *denotative meaning,* coincides with object recognition; a middle-level meaning, *connotative meaning,* refers to the emotional values associated with the object; and a higher-level meaning, *abstract meaning,* refers less to the object than to broader values (Rapoport, 1990b). When you recognize an area as a commercial strip, you experience a denotative meaning. Lower-level meanings include other everyday manifest meanings identifying intended uses of settings (Rapoport, 1990b, p. 221). When you make inferences—such as guessing the likely quality of goods or the friendliness of the merchants in a commercial strip—or evaluative judgments—such as how much you like the appearance of the area—you experience connotative meanings, or what I refer to as *likability.* When you look at the place through "cosmologies, cultural schemata, worldviews, philosophical systems and the sacred," you experience abstract meanings (Rapoport, 1990b, p. 221). This book stresses connotative meanings because of their relevance to shaping urban form and their importance to human behavior. Where people have the capacity to act, connotative meanings affect their behavior, influencing decisions about whether to go somewhere and how to get there.

Lynch (1960) identifies five kinds of elements that give identity to a city: landmarks, paths, districts, edges, and nodes . *Landmarks* are visible reference points. Depending on the scale of reference, they may be large objects such as towers or mountains or, on a local scale, smaller objects such as signs or even a doorknob. *Paths* are channels for movement, such as streets or walkways. *Districts* are larger sections that have some recognizable, common perceived identity, homogeneity, or character (such as Little Italy) distinguishing them from other areas. *Edges* are barriers or boundaries, such as shorelines, rivers, railroad cuts, and walls. *Nodes* are focal points of intensive activity to and from which people may travel, such as city squares. Many studies confirm the stability of these five elements across a variety of populations and cities (Appleyard, 1970; Francescato & Mebane, 1983; Harrison & Howard, 1972; Milgram & Jodelet, 1976). Controlled experiments using a statistical technique called *cluster analysis* also confirm the validity of the five elements (Aragones & Arredondo, 1985; Magnana, 1978). Though some studies question aspects of Lynch's scheme (deJonge, 1977; Gulick, 1963; Klein, 1967; Rapoport, 1977), the differences relate to the sociocultural and physical context of the areas and populations. For example, student maps in Lebanon stress districts, paths between districts, and sociocultural associations as well as the visual form (Gulick, 1963). Maps in Holland tend to stress a dominant path, nodes, and, to a lesser extent, landmarks (deJonge, 1962). Other studies point to other factors: small-scale detail and human activity for children in Berlin; human scale in Birmingham, England; and the congruence of form and activity and mode of travel in Boston (Rapoport, 1977, pp. 115-116). Still, studies of the image of cities around the world—Chicago (Saarinen, 1969); Ciudad Guayana, a new town in Venezuela (Appleyard, 1970); Paris (Milgram & Jodelet, 1976); Rome and Milan (Francescato & Mebane, 1983); Tripoli, Lebanon (Gulick, 1963); and Amsterdam, Rotterdam, and the Hague (deJonge, 1962)—confirm Lynch's conclusions about image formation and structure. The structure (the relationships between the imageable elements) contributes to the vividness, clarity, or legibility of the image. Though the images and prominence of various elements vary for different populations and places, the imageable elements, correctly arranged, can heighten the imageability of a city.

For shaping city appearance, however, knowledge about identity and structure (or imageability) is not enough. Lynch (1960) agrees that legibility may be necessary but is not sufficient for a likable environment. Recall that he sees the city image as having three components—identity, structure, and meaning. Humans have feelings and associations, both negative and positive, about their surroundings and the imageable elements. These feelings and meanings define the evaluative image of a city (Figure 1.2).

These feelings and meaning are also central to our perception of and reaction to the environment. Explorations of people's reactions to various kinds of places have repeatedly found evaluation and appearance to be a primary dimension of the environmental experience (Carp, Zawadski, & Shokron, 1976; Harrison & Sarre, 1975; Hinshaw & Allot, 1972). Evaluation and imageability interact. We recall places about which we have strong feelings, and we will more likely have feelings about and appearance with the recalled (imageable) parts of the city (Rapoport, 1970). Meaning heightens imageability, and imageability intensifies meanings. Confirming this, research has found that the most imageable buildings in a city elicit the strongest evaluations both positive and negative (Appleyard, 1976). Other research has found that meaning guides individuals' selection of imageable elements (Harrison & Howard, 1972) and that features associated with evaluation also relate to building imageability (Evans, Smith, & Pezdek, 1982). Public evaluations of imageable elements will

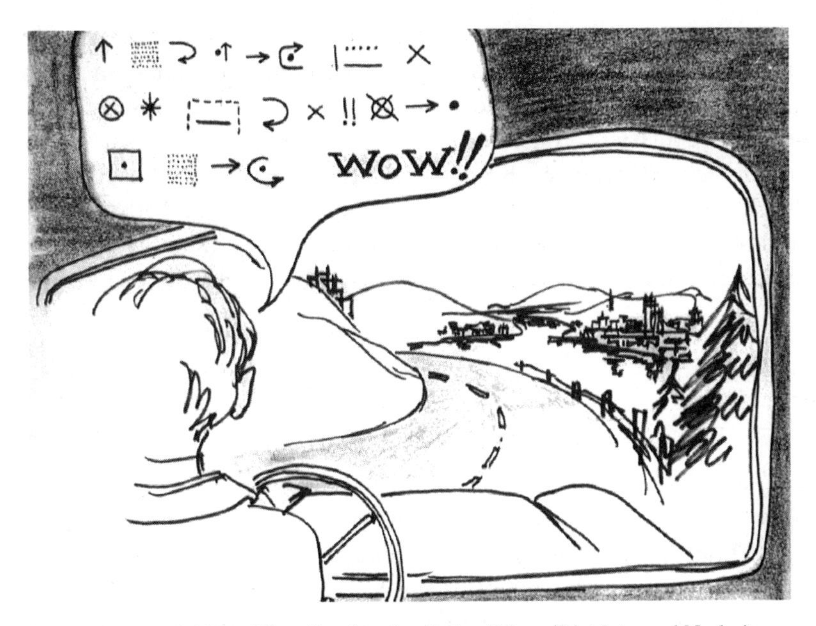

Figure 1.2. Imageability (Clear Landmarks, Paths, Edges, Districts, and Nodes) Is Not Enough; the Evaluative Image Also Depends on Feelings and Meanings
Cartoon courtesy of Oscar Newman.

define the perceived visual quality of the city. If most people like the imageable elements, the city will probably convey an agreeable evaluative image. If they dislike them, the city will convey a disagreeable evaluative image, suggesting a need for changes in appearance.

Although Lynch (1960) recognized the importance of meaning and evaluation, his research emphasizes identity and structure. He felt that people have more consistent perceptions of identity and structure than of meaning. He also felt we cannot easily manipulate meaning through changes in urban form because "vast numbers of people of widely diverse backgrounds" hold many varied images in relation to the same physical form (pp. 8-9). Confronted with possible measurement problems and individual differences, he judged meaning as impractical to study and concentrated on form—identity and structure—separate from meaning. Yet, his research revealed some shared evaluations and meanings. He reported that observers favor "distant panoramas," "the city lights at night," "vegetation or the water," and "upper-class" over "lower-class" areas.

Subsequent research has shown ways to measure environmental evaluation and meaning, and it has shown environmental evaluation as less idiosyncratic than Lynch believed. Various reviews of this research have identified consistent consensus in people's visual preferences in the environment (Kaplan & Kaplan, 1989; Nasar, 1989a; Nasar, 1994; Ulrich, 1983; Wohlwill, 1976). This book goes beyond identity and structure to emphasize the shared evaluative image or likability of a city's visual form and the importance of the evaluative image for city design.

Precedents

To know the meaning conveyed by city form, we need to know how people evaluate the prominent features. This combines what people know about their city (environmental cognition) with how they feel about it (environmental assessment; Evans & Garling, 1991). This approach fits the nonverbal communication approach advocated by Rapoport (1990b). Seeing the built environment as a form of nonverbal communication, Rapoport highlights *pragmatics* as an approach to understanding the process by which the environment functions as a sign. Pragmatics involves examining the effects of nonverbal environmental cues on the individuals "who interpret them as part of their total behavior . . . in a word their meaning" (p. 38).

Several studies have particular relevance to the nonverbal communication approach taken in this book. In 1968, planner Carl Steinitz mapped *denotative meanings*—public knowledge of the city. From a field reconnaissance, he mapped form and activity type, form and activity intensity, and form and activity exposure (or significance). He found the measures of form and activity correlated. From interviews, he obtained public judgments of the type, intensity, and significance of places and activities. Relating this knowledge to the actual form and activities in the city, he found strong similarities with "strikingly few differences" in "which places were best known, how they were identified and described, and for whom they were meaningful" (Steinitz, 1968, pp. 240, 245). Although it is a notable work on meaning, this work overlooked connotative meanings—people's feelings about the places and activities.

Eight years later, Donald Appleyard (1976) a landscape architect and former colleague of Kevin Lynch, reported research in Ciudad Guyana that assessed connotative meanings. He obtained opinions about the attractive and ugly parts of a road, reasons for the opinions, and evaluations of several buildings. He found strong agreement and evidence of the importance of evaluative response, leading him to conclude that "buildings were usually viewed in an evaluative manner" (p. 98) and "both inhabitants and the public at large perceive the urban environment in evaluative terms" (p. 238). Early texts on perceived environmental quality also report studies obtaining evaluative responses to other kinds of places, including shorelines, buildings, a park, waterfalls, and other natural areas (Craik & Zube, 1976; Zube, 1980).

Geographer Peter Gould (1973) took the evaluative responses one step further by putting them in spatial or map form at a national and international scale. He interviewed students about comparative preferences for states in the United States. From the interviews, he constructed evaluative maps of the nation as seen from California, Minnesota, Pennsylvania, and Alabama. The maps reveal shared preferences and differences relating to the location of the observer. Views from several northern states (Figure 1.3) display similar patterns of preference—high preferences for California, with a steady decline moving east (except for Colorado), an increasing preference from the midwest to the northeast, and a decreasing preference to the south. Views from a southern state show the same northern and western peaks, but, unlike the northerners, who tended to lump the south together, the southern students showed a finer discrimination between various southern states. Instead, they tended to lump together much of the north. Using similar methods, Gould also developed evaluative maps for nations in Europe and Africa.

Surely we can develop evaluative maps of cities to use for the analysis and improvement of city appearance. Gould reports some examples (Gould & White, 1974): One investigator used a newspaper poll to construct a preference map for Birmingham, England (Figure 1.4); another one mapped fear in inner-city Philadelphia (Figure 1.5). The Parisian newspaper *L'Aurore* published a map of Manhattan showing safe and unsafe areas for pedestrians during the day and after dark; the *New York Times* responded with a similar map for Paris ("A French View of New York," 1972).

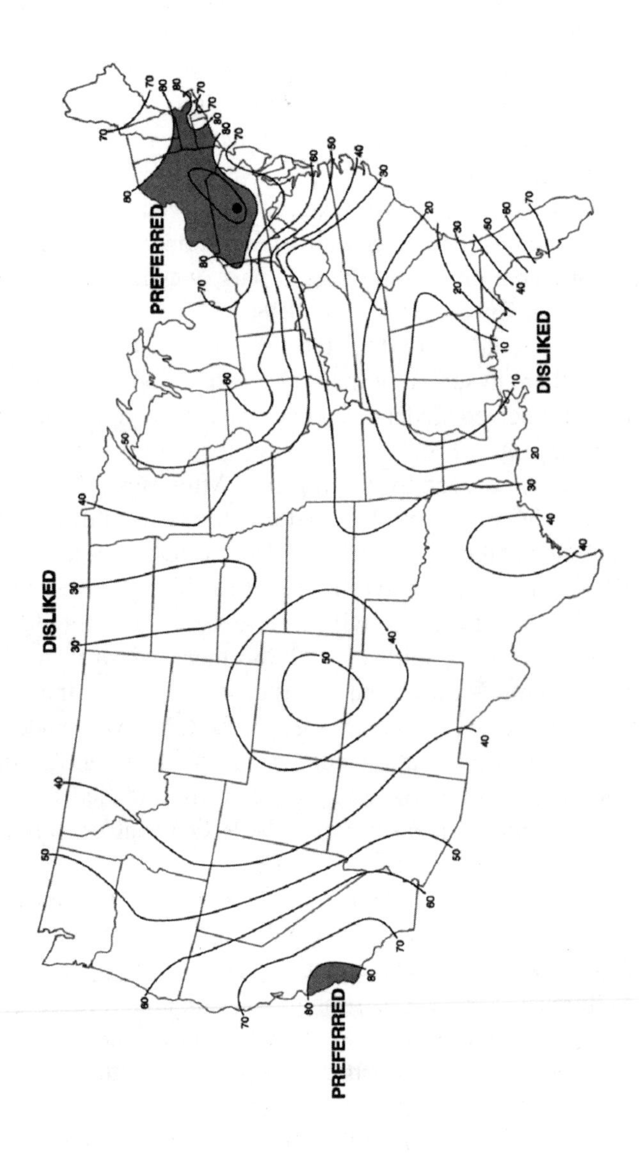

Figure 1.3. Northerners' Preference Map of the U.S.
Low numbers in the south stand for low preferences; high number in the northeast and California stand for high preferences.
SOURCE: Peter Gould (1973); redrawn by David Miller.

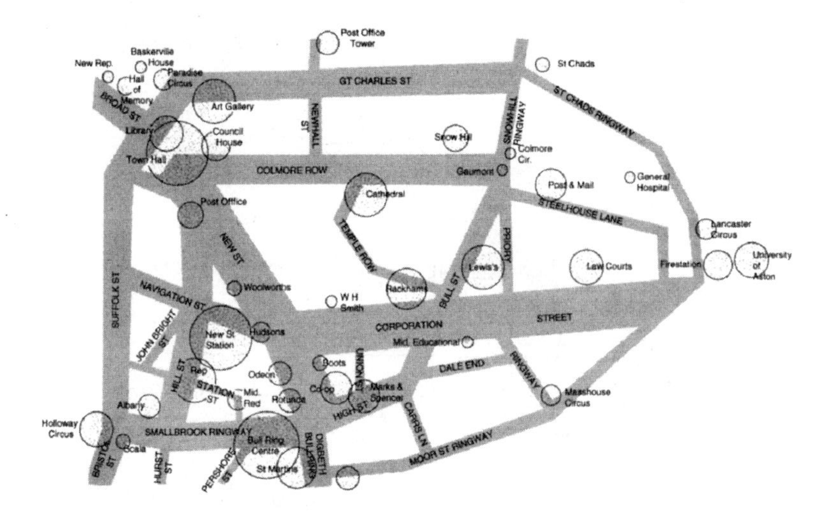

Figure 1.4. Preference Map of Birmingham, England
SOURCE: Gould (1973); courtesy of Peter Gould; redrawn by David Miller.

In more refined research on Paris, psychologist Stanley Milgram obtained several psychological maps of the city (Milgram & Jodelet, 1976). His research team used several tasks to describe the mental map of the city. Capturing denotative meanings, these tasks included (1) hand-drawn maps, (2) associations to map elements, (3) a photo recognition task, and (4) having people say where they would wait to meet someone to maximize the chance of encountering him or her. From the hand-drawn maps and meeting-place question, the researchers identified the most frequently cited elements. From associations to elements on the maps, the researchers found the embeddedness or number of links to other locations. The photo recognition task allowed the researchers to identify the familiar and unknown parts of the city. They also looked at connotative meanings. For example, they asked (1) where the rich and the poor live, (2) where the dangerous areas are, (3) where the snobby Paris is, (4) where one would move if one became wealthy, (5) where the friendlier (more relaxed atmosphere) is, (6) where one would walk if one had one last chance to walk through the city, and (7) what areas one liked best. The researchers found strong agreement in evaluative response. Most respondents agreed on the top-ranked areas. For example, 70%

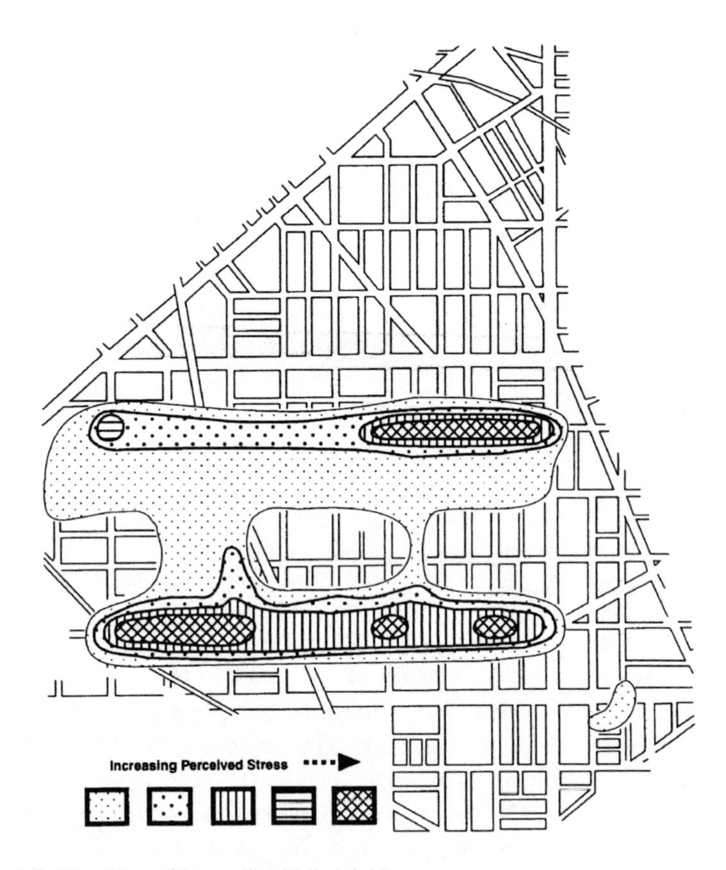

Figure 1.5. Fear Map of Inner-City Philadelphia
SOURCE: Gould (1973); courtesy of Peter Gould; redrawn by David Miller.

identified one area as the best liked, and 88% agreed on one area as "the Paris of the rich." When asked where they would take a last walk, respondents most often selected four streets and, to a lesser extent, an additional four streets, though Paris has roughly 3,500 streets. Milgram presents some of the findings in map form, including maps of the perceived rich and poor areas, the last walk areas, and the perceived dangerous areas (Figure 1.6). Milgram emphasizes the psychological component of the maps. For urban design, we need information that also considers how the evaluative image relates to the physical characteristics of the areas. I take Gould's (Gould & White, 1973) idea of regional preference maps, Milgram's idea of psychological maps of the city, and

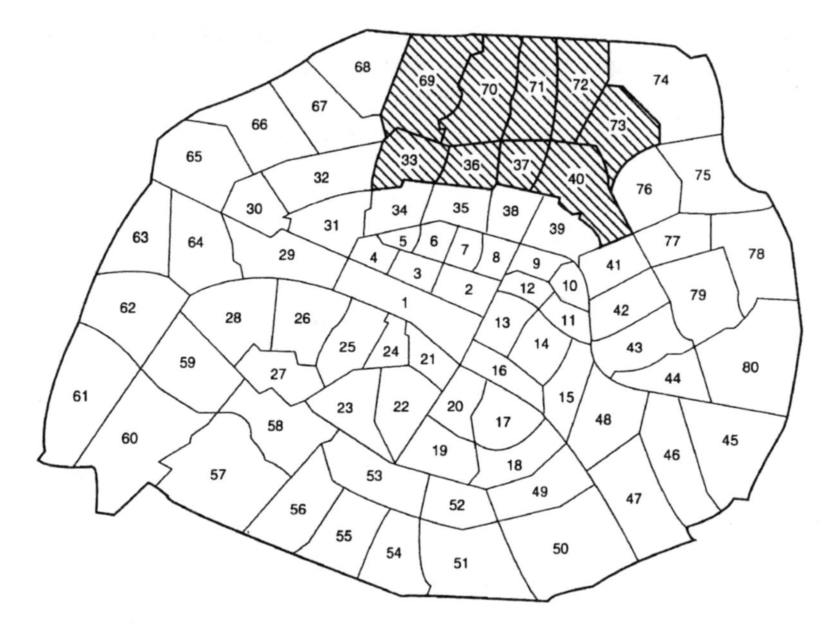

Figure 1.6. Perceived Areas of Danger in Paris
SOURCE: Milgram and Jodelet (1976); redrawn by David Miller.

Lynch's (1960) approach of linking psychological information to urban form, and apply them to the study of city appearance.

In two U.S. cities—Knoxville and Chattanooga, Tennessee—we interviewed 300 residents and visitors. We looked at visitors as well as residents because of likely differences between insider (resident) and outsider (visitor) meaning (Brower, 1988). We asked respondents about the areas they liked and the areas they disliked visually, and their reasons. From each interview we prepared an evaluative map of the city. We then overlaid these maps to produce for each city and for residents and visitors separately a composite map representing the evaluative image of the city. In considering mental maps such as these, bear in mind that the map-in-the-head idea is only an analogy or metaphor, in that people act as if they have something like a map in their head. The process—mapping—differs from the product—the internal mental structure, often called a *cognitive* or *mental map* (Downs, 1976, p. 67). The mental map has incomplete, distorted, and simplified information. The evaluative maps suggest associations with city structure and experience, and they indicate likability associated with five features—naturalness, upkeep, openness, order, and historical significance. They provide a basis for a visual plan for guiding the future appearance of a city. The maps show the identity, location, and likability of visual features, and they explain the basis for the evaluations.

2

Measuring Community Appearance

Research on the evaluative image of the city might have little practical value if design professionals shared the values of the public and delivered those values in designs and plans. This has not been the case. Outsiders tend to evaluate places differently from insiders, who are familiar with the area (Brower, 1988). For example, in the 1960s, design professionals saw a slum in Boston's West End and removed it. The residents, familiar with the area, had experienced a vital neighborhood (Gans, 1962). Similar misjudgments have occurred elsewhere. Designers misjudged another area as a slum, in spite of its good maintenance, because they despised the materials residents used for renovations (Sauer, 1972). For a new British town, planners and residents had opposite evaluations: The residents liked its features, and the planners did not (Bishop, 1984). In another case, residents had negative responses to a feasibility study for a development praised by architects and the architectural press. The residents described it as a "prison," a "concentration camp," "depressing," and "claustrophobic" (Rapoport, 1990b, p. 232) Observers have noted the disregard many designers (outsiders) have for popular (insider) values (Blake, 1974; Gans, 1974). For example, in a direct attack on user meanings, designers and company executives tried to furnish the interior of a building by a well-known designer (Saarinen) to prevent a "kewpie doll atmosphere." This provoked a suit from the workers, who preferred a different appearance (Rapoport, 1990b, p. 21).

Figure 2.1. Example of the Type of Style Most Liked by Architects and Least Liked by Nonarchitects
Photograph by Jack L. Nasar.

For more than 60 years, planners and designers have attacked the suburban landscape. Yet the public around the world continues to evaluate it favorably as the ideal (Michelson, 1976; Rapoport, 1993).

Research confirms the anecdotal accounts of differences. It consistently finds architects as having building preferences and meanings that differ from those of the public, and it finds the architects as misjudging public preferences and meanings (Devlin & Nasar, 1989; Groat, 1982; Kang, 1990; Lee, 1982; Nasar, 1989b; Nasar & Kang, 1989). What architects like, the public dislikes, and what the public likes, architects dislike. They see different meanings in the same buildings. Consider a study in which we had architects and other nondesign professionals evaluate high-style and popular-style homes (Devlin & Nasar, 1989). The architects liked the high style best, and the nondesigners liked the popular style best (Figures 2.1 and 2.2).

A related study on design review found that design review commissions see different meanings than the public they supposedly serve. It shows a low correlation between traditional design review methods and

Figure 2.2. Example of the Type of Style Least Liked by Architects and Most Liked by Nonarchitects
Photograph by Jack L. Nasar.

public preferences (Stamps, 1992c, 1997). Research on design competitions also suggests that architects may not lead public taste. In a study of public and architect evaluations of design competition winners and losers over a 100-year period, I found that both groups preferred the losers to the winners (Nasar, 1998). They also described more losers than winners as the better design. Examinations of public reactions to modern architecture, even after more than 60 years, indicate a public distaste for the form (Nasar, 1994). The differences between designers and the public can result in solutions that are unappealing to the public. In contrast, two findings suggest that research on popular preferences can guide design. In each one, popular appraisals of designs agreed with subsequent popular appraisals of completed buildings (Nasar, 1998; Stamps, 1997).

The mismatch assumes greater importance because of the importance of environmental meaning and evaluative response. A large body of research indicates the significance of environmental meaning to people (Rapoport, 1990b). Humans naturally notice things and draw meanings from them. According to some theorists, this ability has a basis in our evolution (Kaplan & Kaplan, 1989; Mars, 1996; Ulrich, 1983): To survive, early humans had to notice and evaluate the meaning of events that might have benefited or threatened their well-being. Individuals lacking that ability did not survive and pass on their genes. Over thousand of years, evolution selected individuals who noticed, drew meaning, evaluated things around them, and then acted accordingly.

Studies consistently find meaning and appearance as the major components of human response to buildings, houses, neighborhood, and community. Several studies highlight the importance of meaning to people: Research on vernacular design in Greece found meaning as the most important aspect of response; research on student satisfaction with resident halls found the evaluative meaning (the character and feel) as most important; and research on the image of Phoenix, Arizona, found that respondents stressed affect and associations—evaluative and social meanings (Rapoport, 1990b, pp. 234, 30, 14). Other research confirms the importance of appearance, affect, and meaning. A study of buildings found a space-evaluative factor made up of scales such as *pleasing, delightful, beautiful,* and *interesting* as a dominant aspect of response (Hershberger, 1969). Another study found judgments of community satisfaction primarily related to aesthetic factors (Lansing, Marans, & Zehner, 1970). A study of responses to British urban places found that most responses consisted of emotional appraisals (Burgess, 1978). Other research found appearance and evaluative variables as explaining the highest proportion of the variance of ratings of residential quality (Carp, Zawadski, & Shokron, 1976). Though some of these studies may suffer from a bias in that they relate one verbal measure to another, research avoiding this bias has also found evaluation as the most prominent factor in responses to residential scenes (Horayangkura, 1978). We can also see the importance of appearance in behavior. For example, low-income residents in self-built prefabricated dwellings first decorated the dwelling exterior before adding insulation or security features, even though the area had problems with climate and crime (Nasar & de Nivia, 1987). Similarly, when officials relocated residents due to flooding from the Aswan Dam, the residents first changed exterior decorations and colors of their new dwellings (Rapoport, 1990b, pp. 92-93). The German city of Munster, rebuilding after destruction from World War II, maintained the same appearance as before (Gleye, 1983), and the Ohio town of

Xenia, rebuilding after a tornado destroyed the downtown, copied the earlier form of development (Francaviglia, 1978). The evidence consistently shows the centrality of visual character and meaning to human experience.

How do we arrive at guidelines for decisions on community appearance? Efforts have come from two directions: the historical or philosophical approach and the scientific approach.

Speculative Versus Empirical Approaches

The Speculative Approaches: Historiography and Speculative Philosophy of History

Two subdisciplines of history try to draw patterns from the historical record: historiography and speculative philosophy (Simonton, 1990). Historiography tries to establish and convey facts to the reader. It stresses the particulars, such as specific persons, dates, and places. As such, it is *ideographic.* Speculative philosophy of history, or what some writers refer to as *architectural theory* or *speculative aesthetics* (Groat & Despres, 1990; Lang, 1987), tries to support conclusions from an analysis of patterns. This kind of approach might profile a select set of "masterpieces" and "master" designers. The analysis tries to build an argument in support of the author's theory, a normative statement for a particular aesthetic. It describes the way things ought to be rather than the way they are. Furthermore, it stresses a limited set of monumental buildings (such as churches, institutions, or palaces) and a limited set of architectural "hero" figures. The narrow view makes the results unrepresentative. The monumental buildings occupy less than 2% of the built environment; the architectural heroes do not represent most practicing architects or, for that matter, the many other people who influence development (Rapoport, 1990b). The approaches also suffer from a narrow emphasis on architecture as an artifact and art. They overlook important relations between the building and human behavior, between the various parts of the building, and between the building and the larger setting (Rapoport, 1990b). An art object (in the high-culture sense) does not need to function or work technically for users. It only has to satisfy judgments of its artistic merit by critics and other artists.

A change in emphasis from buildings to the full environment and from architectural heroes to everyone involved in shaping the environment would ease some problems of representativeness, but the approaches would still suffer from threats to the validity of the evidence. To present an interesting narrative or to argue for a particular theory, historians and theorists arrange and choose which facts to emphasize and which to omit. The argument stands less on objective criteria than on its subjective persuasiveness for an ideology (Simonton, 1984). Though these approaches have considerable value for understanding the nature of individual experience from inside and for bringing attention to inequalities, they show little evidence of advance in knowledge for built environments (Rapoport, 1990b). We cannot build a knowledge base for guiding community appearance on the foundation of such opinions.

To overcome these problems, the approaches must change "from an art history metaphor to a science metaphor" (Rapoport, 1990a, p. 58). Without necessarily dropping historical or philosophical inquiry, the change calls attention to the applied value of the special kind of knowledge obtained through science. The scientific approach builds a knowledge base to guide future actions. It builds and tests theory that helps us understand, predict, and act on things. It can provide the basis for design guidelines on community appearance.

The Scientific or Empirical Approach

Can we scientifically study something as apparently subjective or qualitative as community appearance, evaluation, and meaning? Yes. It fits in the realm of social science. Social scientists have developed methods to study humans and subjective experience. Because social science relies on questions and observations instead of technical equipment and looks at humans and subjective experience, some people view the results as contestable.[1] They incorrectly view the methods as ordinary, and they question the validity of the findings relative to their own commonsense experience and guesses. When the research addresses deeply felt beliefs and values, people may not readily accept "facts" that contradict their beliefs and values. In rejecting the findings, they overlook an important distinction between ordinary knowing and scientific knowing.

Scientific knowing differs from commonsense guesses or ordinary knowing in ways that make the scientific knowing a more valid basis for action. Like social science knowing, ordinary knowing relies on observation to put together explanations and to evaluate their validity and relevance, but the evidence in ordinary knowing has more biases. Without realizing it, individuals often seek and give more weight to confirmations of their beliefs or wishes and ignore inconsistencies in beliefs (Feather, 1964). They may use biased heuristics in making judgments (Nisbett, Krantz, Jepson, & Kunda, 1983; Tversky & Kahneman, 1974). They may rely on opinions of authorities or peer groups, neither of which may be accurate (Ross, Amabile, & Steinmetz, 1977). They may have biased observations and inaccurate recall of past experiences (Loftus & Hoffman, 1989; Snyder & Swann, 1978). In each case, the biases tend to support the beliefs or naive hypotheses of ordinary knowing (Snyder & Swann, 1978).

In contrast, scientific knowledge derives from the scientist's skeptical or conservative position toward the hypothesis. Scientists try to construct hypotheses so that experimental data can be unambiguously interpreted. They rely on systematic empirical observations to test a claim. They look for biases to determine the relative merits of various explanations. Good science does not involve seeking to prove a hypothesis or theory. It involves ruling out rival hypotheses and evaluating the degree to which data agree with a hypothesis. Ordinary observers may also look for biases, but they do so to a lesser degree than the scientist does. Scientists, aware of research on biases, systematically study ways to avoid biases in examining hypotheses, and they use that information in the design and evaluation of research. Scientists may have biases and a vested interest in a finding, but the structure of scientific inquiry helps make science "self-correcting, progressive, and cumulative" (Rapoport, 1990a, p. 64). Scientists make the details of their work—the questions, assumptions, and methods—explicit, public, and open to replication and falsification. The peer review process has other scientists review the merits of the work before publication. Afterward, other researchers can replicate or run variations on a study to test the conclusions. If others feel that they have a better way to manipulate or measure a variable, they can conduct another test. In addition, the public nature of science allows investigators to place and evaluate the findings in the context of other research and theories. All this contributes to building a scientific knowledge base.

Scientists examine theoretical hypotheses in a systematic way, attempting to avoid biases. They define constructs of theoretical interest for study and develop observable indicators to measure the constructs.[2] They define and sample units for statistical analysis. They examine the relations among the constructs, and they use statistical analyses to determine causal connections or associations between the constructs (Judd, Smith, & Kidder, 1991; Simonton, 1984). For example, suppose we want to test the theory that naturalness has aesthetic value. We would have to develop operational definitions of the constructs *naturalness* and *aesthetic value*. These definitions tell us how to measure the construct. They give a concrete and specific definition of the measure that allows others to repeat the measure. The operational measures serve as "indicators from which inferences can be made about the higher level constructs" (Rapoport, 1993, p. 9). We would then measure the constructs and analyze the results.

Two kinds of relations between variables serve as hypotheses: (1) the relations between constructs and observable indicators of them, and (2) the relations between the constructs of interest. The first checks whether the measures represent valid indicators of the constructs, such as naturalness and aesthetic value. The second checks whether a construct (such as naturalness) produces another construct (aesthetic value) for a certain population (such as residents of a neighborhood) in a condition (such as a residential street).

This approach allows investigators to develop an empirically measurable test of the phenomenon. Still, scientists must make interpretations and inferences from the findings; these steps away from the data may have flaws. Studies also may lack consistency or clarity in the way they define or operationalize various constructs. For example, for evaluation of naturalness, different studies may measure preference, attractiveness, and satisfaction and they may use different kinds of natural stimuli. As a result, the studies may not capture the same phenomena. Because any one method has potential biases, researchers recommend the use of multiple methods (Campbell & Fiske, 1959). Convergent findings across different methods are more likely stable.

Through research design and methods, scientists try to control or remove confounding influences other than the experimental variables and try to reflect the actual situations to which the results might apply (Campbell & Stanley, 1963; Cook & Campbell, 1979). With this in mind, scientists consider several factors in evaluating the soundness of a study.

They look at the degree to which the measures of the variables accurately reflect the construct (*construct validity*). They look at the degree to which the measure obtains consistent results in which the observed conditions have not changed (*reliability*). They look at whether or not changes in the independent variable caused changes in the dependent variable (*internal validity*). They look at the extent to which the findings generalize to other situations, populations, and measures (*external* or *ecological validity*). Several books give more detailed discussion of validity and reliability (e.g., Campbell & Stanley, 1963; Cook & Campbell, 1979).

In sum, social science uses systematic methods and a skeptical outlook to build a knowledge base. Science examines hypotheses central to a theory. The theory should address some important unexplained phenomenon, and it must include hypotheses about measurement and about the relationship between variables.

> When it comes to knowing (at the cognitive level) and being able to communicate and use such knowledge, the thing we call science is by far the best, most elegant, most powerful and most successful way so far developed to achieve *well-founded, reliable knowledge.* (Rapoport, 1990a, p. 60)

Communities need this kind of knowledge to guide their decisions on community appearance.

Domain of Study

The evaluative image represents a psychological construct that involves subjective assessments of feelings about the environment. This suggests that the evaluative image contains two kinds of variables: visual aspects of city form and human evaluative responses. The visual features are the independent variables, and human evaluative responses are the dependent variables. For urban design, a place has a favorable evaluative image or looks good if enough ordinary people (those experiencing the place on a regular basis, rather than experts) say so. Thus, urban designers want to know what noticeable features of the visual environment are associated with favorable meanings or likability in the evaluative image.

The Relevant Physical Environment and Human Responses

For the physical environment, those aspects of development visible from public space—from "the exterior of . . buildings outward" have particular relevance for community appearance (Shirvani, 1985, p. 6). Whether a development occupies private or public land or uses private or public resources, the exterior stands as a public object. U.S. federal and state courts have held developments subject to public controls such as building codes, sign ordinances, and design review (Pearlman, 1988). Public spaces and the parts of the community visible from public spaces represent the primary areas of concern for city design. Though these places have many features, the noticeable features have the most relevance. We need to consider the prominent perceptual features. Places consist of fixed features, semifixed features, and nonfixed features (Rapoport, 1993). Fixed features are permanent or slowly changing. Nonfixed features, such as people and animals, are constantly changing. Semifixed features are changeable elements such as signs, billboards, plants, and decorations. They are relatively easy to alter, and they may have a greater effect in communicating identity and meaning. For community appearance, then, urban design has a special interest in the noticeable semifixed features in or visible from public spaces.

For human response, I replace the term *aesthetic response* with *evaluative response* to convey a broader meaning and to eliminate associations with artistic expression. Traditional definitions of aesthetics often refer to the perception of beauty in the arts and imply extreme and intense feelings such as the sublime (Lang, 1987, p. 179). Some critics question the use of such a definition. They suggest that the identification of high-culture art with culture leads to a groundless distinction between high (or serious) art and low art or entertainment (Gans, 1974; Peckham, 1979). One critic defines art contextually as objects occurring in settings, such as museums, that house art (Peckham, 1976). Psychologist Joachim F. Wohlwill (1974) notes that the traditional definition of aesthetics overlooks smaller changes in emotional response that people regularly experience as they move through their communities. Following these critiques, I have broadened the definition of aesthetic response to include these less extreme affective responses. I also have broadened the definition to include meanings associated with places. Unlike art, literature, and music, which the observer can choose whether or not to experience, city design does not afford the observer such a choice. In their daily activities, people must pass through and experience public parts of the city environment. As a result, appraisals of city appearance should replace an emphasis on city form as an object of art with evaluations by people experiencing the city form.

The evaluative image does this. It refers to favorable emotions and meanings experienced in relation to the environment. Though pleasure represents an important component of evaluative meaning, the image also has other dimensions. Using a variety of research strategies and measures, psychologists James Russell and Larry Ward found four dimensions—*pleasantness, arousing, exciting,* and *relaxing* (Russell & Snodgrass, 1989; Ward & Russell, 1981). Though other studies have examined dimensions of meaning, Russell and Ward's research has most relevance to environmental assessment. Unlike some studies (Heise, 1970; Osgood, 1971; Osgood, Suci, & Tannenbaum, 1957), Russell and Ward stress responses to the physical environment; unlike other studies (Canter, 1969; Hershberger, 1969; Kuller, 1972), Russell and Ward do not confound judgments of environmental features with emotional appraisals. They focus on dimensions of emotional meaning for the physical environment. Figure 2.3 shows the circular ordering of the affective appraisals and emotional reactions.

Pleasure is a purely evaluative dimension. Arousal is independent of the evaluative dimension. Excitement and relaxation involve mixtures of evaluation and arousal. People experience exciting places as more pleasant and arousing than boring ones; they experience relaxing places as more pleasant but less arousing than distressing ones. Subsequent research has confirmed the relevance of these dimensions of emotional meanings for appraisals of urban scenes (Hanyu, 1993, 1995; Nasar, 1988a).

Beyond these affective responses, humans also experience connotative meanings. You might look at an area and judge it as prestigious or unsafe, or you might guess that the residents are friendly. When you make inferences like these, you infer associational or connotative meanings. Research shows that people do infer such meanings about places (Rapoport, 1977, pp. 65-80). From environmental cues, people can make inferences about prestige, social position, class, or status (Cherulnik & Wilderman, 1986; Duncan, 1973; Lansing, Marans, and Zehner, 1970;

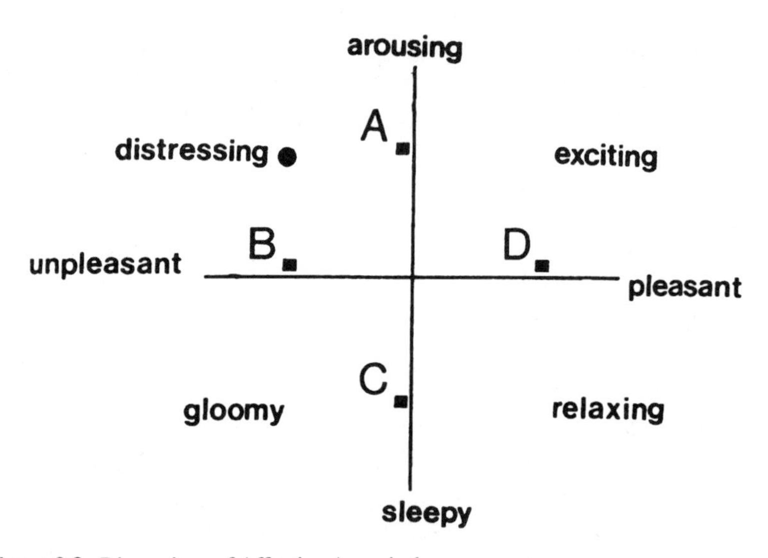

Figure 2.3. Dimensions of Affective Appraisals
SOURCE: Russell & Snodgrass (1989); courtesy of James Russell.

Nasar, 1989b; Royse, 1969), personality, use, identity (Rapoport, 1993; Sadalla, Verschure, & Burroughs, 1987), friendliness (Nasar, 1989b; Sanoff & Sawney, 1972), neighborliness and homogeneity of neighbors (Lansing et al., 1970, suitability to a campus and livability (Nasar & Kang, 1989), safety (Nasar & Jones, 1997), privacy, territoriality, and, for potential burglars, criminalizability (Brown & Altman, 1983; Newman, 1972). These meanings may influence the emotional response and behaviors in relation to an area, and they may play an important role in the evaluative image (Harrison & Sarre, 1975).

Importance of Community Appearance

The measurement of the evaluative image, meaning, and community appearance has importance for creating an objective basis for decisions and for policy reasons. In contrast to the conventional wisdom that beauty is in the eye of the beholder, research shows strong consistencies in what people like and dislike in the environment. More than a century ago, psychologist Fechner (1876) showed that researchers could study visual preferences scientifically to reveal and quantify patterns of preference. More recent study has confirmed strong consensus in environmental preference, providing a quantitative basis for design decisions (see reviews by Kaplan & Kaplan, 1989; Nasar, 1988c; and Purcell, 1986). Meanings may vary with sociocultural conditions, but residents in an area and sociocultural group will likely have shared cultural meanings in relation to their environment (Rapoport, 1977). Research has also confirmed similar preferences across cultures (Hull & Revell, 1989; Nasar, 1984; Ulrich, 1993). Though individuals and groups may have idiosyncratic preferences, they generally agree on certain components that make for a desirable urban form. These findings suggest that beauty in the environment is less qualitative and subjective than many people think it is.

Studies of the evaluative image and meanings can provide valid, reliable, and useful information for the planning, design, and management of desirable surroundings. As landscape architect Erv Zube (1980) notes,

> Modifications in the environment will occur sometimes under the aegis of federal, state or local policies. Evaluation studies provide a means for assessing the efficacy of those policies as evidenced by the success or failure [of the developments that result] based on empirical data rather than on guesses and intuition. (p. 1)

Public policy in the United States has also acknowledged the importance of visual quality and meaning in the environment. Aesthetics and related terms such as *beauty, compatibility,* and *harmony* appear in federal, state, and local planning guidelines. The National Environmental Policy Act and the Coastal Zone Management Act mandate consideration of aesthetic variables. Federal and state courts grant aesthetics alone as an adequate basis for design controls (Pearlman, 1988), and most communities use various financial, administrative, and regulatory techniques to control appearance (Shirvani, 1985). Sign ordinances and guides for them are widespread (Ewald & Mandelker, 1977). One survey of local U.S. planning departments found that 93% of communities with populations over 100,000 and 83% of cities with populations over 10,000 use some form of design review to control appearance (Lightner, 1993). Another study of more than 1,000 cities indicates that more than 90%

have controls for appearance (International City Management Association, 1984).

Although many questions remain, community appearance may suffer if we continue to treat visual quality as a matter of taste. For urban design, we need to try to understand the principles underlying evaluative response and to transform those principles into guidelines for shaping urban form.

A Theoretical Framework

Although research in environmental preferences often takes a stimulus-response form that suggests a certain kind of determinism, I believe that the preferences conform to an interactional perspective (Moore, 1989). The evaluative response arises from the person, the environment, and the interaction between the two. Cognitive processes represent important mediating variables in human evaluative response.

This means that evaluative response and meanings have probabilistic relationships to physical attributes of the environment (Brunswik, 1956). Humans may have a variety of evaluative responses to any environment. Given a set of circumstances (a point in time, a specific group of humans, certain affective states and intentions), an evaluative response has probabilistic relationships to environmental perception and cognition. Perception involves information in the environment picked up by the observer. Cognition involves some internal processing of the information picked up through perception. Perception and cognition, in turn, have probabilistic relationships to one another and to the physical character of the built environment. The probabilities result from the ongoing interaction between individuals (biological, personality, sociocultural factors, and goals) and the environment. Because of shared biology, culture, and environment, humans will exhibit some agreement in their evaluative response. This model suggests two broad components of evaluative response—perceptual and cognitive—and two kinds of environmental variables—formal and symbolic.

First, humans can experience emotion independent of and before conscious perception (Zajonc, 1984). This kind of evaluative response represents a rapid initial response to gross environmental characteristics, called *preferenda.* It precedes and occurs independently of recognition, comprehension, or cognition. It is a direct response to *formal* variables. These variables have to do with the structure of forms. They include physical properties and relationships such as shapes, proportions, rhythms, scale, color, illumination, shadowing, geometry, hierarchy, systems of spatial relations, complexity, incongruity, ambiguity, surprise, novelty, and order (Groat & Despres, 1990; Lang, 1987; Wohlwill, 1976). We respond to them "for their own sake" (Lang, 1987, p. 187).

Second, there has been widespread agreement and evidence that cognition can change affect (Lazarus, 1984). Such cognition need not involve rational calculation but can include such things as categorization or inference without conscious thought (Kaplan & Kaplan, 1989). For the environment, such cognitive influences may take on greater importance than cognition-free affect. Because we inhabit the environment and have to navigate through it, we need to make sense of it (Kaplan & Kaplan, 1989). *Symbolic* or *content* variables take on importance in helping us make sense. They have to do with the meanings of the forms. A dictionary definition of *symbolic* refers to it as "a sign by which one knows or infers a thing," or "to throw or put together" (Neilson, Knott, & Carhard, 1960, p. 2555). It throws together the object and the observer's experience and associations. Beyond the experience of pure form, humans experience the environment through mediating variables. These variables relate to the environment but reflect the individual's internal representation of and associations with it (Moore, 1989). Recall that place meanings can take several forms. *Denotative* meanings refer to judgments of what the place is, and *connotative* meanings refer to inferences about the quality and character of the place and its users. Evaluative responses to content involve both kinds of meanings. They relate to the parts and relationships, an individual's recognition of types, and objects classified as part of a formal structure (Groat & Despres, 1990; Norberg-Schulz, 1965). From the point of view of Gestalt psychologists, people naturally organize the parts to make them more simple and coherent (Koffka, 1935). The whole does not equal the sum of the parts. Rather, observers use laws of organization—such as proximity, continuation, and closure—to create good form from the parts (Coren & Girgus, 1980). Other factors may also organize what we see (Gibson, 1979; Neisser, 1976). In contrast to the appreciation of formal features for their own sake, the evaluation of content depends on a more extensive cognitive process. The individual must recognize the denotative meaning (or the content of a formal structure) and infer connotative meanings about it.

Judgments of formal and content features reflect an interaction between the two kinds of variables. Content classifications depend on formal features, and perceptions of the formal features depend on content. Content classifications involve internal comparisons between the object's formal features and systems of formal features. People recognize architectural styles as content categories because the examples of the style have a recognizable system of formal features (Devlin & Nasar, 1989; Espe, 1981). Content may influence perceptions of formal variables because the content organizes experience. For example, judgments of order may depend on an observer's understanding of the content category. Thus, we found that architects judge high-style houses as more meaningful and higher in novelty, complexity, coherence, and clarity, whereas nonarchitects give higher scores to the popular styles (Devlin & Nasar, 1989). The "objective" appraisals of formal features depend on each group's recognition and comprehension of the styles.

Because the evaluative image results from a two-way process between the observer and the cityscape, we can improve the evaluative image by shaping the observer or shaping the city form. We can educate observers to notice and evaluate things differently. Museums try to do this by providing text or tape-recorded material to explain the art; cities try to do this by designating historic districts or buildings and providing plaques and fliers with information about them. Certainly education can shape experience. Research shows that architectural students' evaluation of architecture changes over the course of their educational experience (Purcell, 1995; Purcell & Nasar, 1992; Wilson & Canter, 1990). Changes in cognitive set influences evaluative response (Leff, 1978; Leff, Gordon, & Ferguson, 1974). We have observed changes in evaluative response associated with different labels (public housing versus condominiums) assigned to the same buildings (Nasar & Julian, 1985). Other research confirms the effects of predispositions on environmental evaluation (Gifford, 1980).

To some extent, signs and information may enhance people's appreciation of their surroundings. For public places such as a city or neighborhood, however, changes in the physical form of the environment can have more direct, widespread, and lasting effects. By shaping the physical and spatial form of our cities (Shirvani, 1985), urban design affects the experience of many observers.[3] For urban design, we must learn how to shape the future meanings of our cities so that humans enjoy the result.

Notes

1. My discussion of ordinary and scientific knowing draws from and expands on a discussion by Judd, Smith, and Kidder (1991).
2. This rests on the assumption that although measures may be imperfect, one can develop measures for all constructs of interest.
3. I use the term *urban* (community or city) *design* to refer to an activity that takes place at many scales, for a variety of land uses, and at various intensities of development. It includes many kinds of projects such as large-scale private-sector development, public conservation of the environmental quality of communities, and low-cost neighborhood improvements by citizens (Appleyard, 1982). It also includes many kinds of activities such as the design and the development of policies and design controls for a large city's central business district, a small rural commercial strip, a residential neighborhood, a mixed-use development, an industrial park, or a retail sign ordinance.

3

Two Cities

I have argued that community appearance affects our experience of the city, evoking emotions and inferences that influence spatial behavior, and that it also remains a paramount concern to the public. To understand the evaluative image of a city, we talked with people about their impressions of their city. Doing this allowed us to test the idea of the evaluative image, learn the features associated with it, and derive design guidelines for city appearance. Professionals who shape city form can benefit from attending to and using the evaluative images. They can also gain from using appropriate research techniques to discover the images. By assessing the evaluative image, I believe we can start to develop appropriate techniques.

I studied the evaluative image of two cities: Knoxville and Chattanooga, Tennessee. This chapter considers the physical form and public evaluative image of these two cities.

In the spring of 1978, organizers of the Knoxville International Energy Exposition created a Community Appearance Committee to assess and promote the improvement of community appearance. In response to their request to the School of Architecture at the University of Tennessee for assistance in this endeavor, I agreed to help develop a systematic inventory of treatable community appearance problems. Typically, visual inventories center on visual form independent of any assessment of public reactions to those forms. They may highlight visual features, such as enclosures, vistas, or visual sequences. To ensure that the public notices and appreciates visual improvements, we need to find out how ordinary persons react to the visual form. Rather than speculate on the city's appearance, I sought public opinions. Students and I interviewed a diverse and representative sample of 160 residents and 120 visitors in Knoxville. We selected residents at random by phone numbers, and we contacted visitors at more than 30 hotels and motels throughout the city. Over 90% of the individuals contacted agreed to participate. We asked respondents to identify up to five areas they liked visually and up to five areas they disliked visually. We probed to identify the boundaries of each area. If the respondents did not know a particular street or edge boundary, we reminded them that they could refer to other elements (such as a retail establishment), from which we would determine the actual boundary. Once individuals had defined the areas, we asked them for the physical features that accounted for their evaluations. Residents stated their responses orally in the phone interviews. Visitors responded orally and in writing on city maps in the in-person interviews.

Chapter 5, "Evaluating the Method," presents a more detailed description and evaluation of the method. The relatively small size of the samples may prevent us from saying that we obtained true public evaluative images. But the internal consistencies in responses clearly point to the presence of group evaluative images and the possibility for research to uncover those images. Furthermore, the agreement of the components of likability with other research on environmental preference indicates the stability of our findings.

Our stress on visual features may have reduced comments on nonvisual features and possibly levels of activity. We emphasized the visual experience because of its prominence and because of the city request for appearance guidelines. A more comprehensive study might explicitly explore different sense modalities and physical and sociocultural attributes (Rapoport, 1993). To include these other elements, one could ask about the character or ambiance instead of visual quality.

We examined the visitor as well as the resident evaluative images for both theoretical and practical reasons. Because these two groups regularly experience different parts of the city and interact with the environment differently, they may develop distinctly different impressions of a city. Visitors tend to experience the city as observers responding to first impressions, whereas residents tend to respond as participants (Brower, 1988). We would expect residents to show a more detailed knowledge of the city—citing more landmarks, paths, edges, nodes, and districts and making finer discriminations between areas—than the visitors. The two groups might use different criteria and have different needs, wants, expectations, and meanings (Brower, 1988). Residents may have adapted to or filtered out features and activities to which the newcomer attends (Wohlwill & Kohn, 1973). Preference may also relate to familiarity and frequency of exposure, leading to different patterns of preference between the two groups (Bornstein, 1989; Moreland & Zajonc, 1977, 1979). From a practical standpoint, the visitor evaluative image has importance to the local economy and to future growth through tourism. In Knoxville, leaders wanted to improve the city's image for the expected 60,000 visitors a week to the International Exposition and for the thousands of visitors to the University of Tennessee, the Tennessee Valley Authority, Oak Ridge, the Smoky Mountains, and the Dogwood Festival. In 1986, visitors spent almost $250 million in Knox County (Knoxville/Knox County, 1988). Tourism can maintain, diversify, or bolster local economies. In Chattanooga, the other city for which we developed evaluative maps, tourism expanded 528% between 1958 and 1978. In 1978, visitors to the area (Hamilton County) spent $188 million (Livingood, 1981). To attract tourists, residents, business, and growth, communities want to convey a favorable impression to visitors. Knowledge gleaned from visitor evaluative maps can help them do that.

From the interviews, we produced 160 resident evaluative maps and 120 visitor evaluative maps, one for each respondent. Though the maps are not the individuals' actual evaluative images, they hold clues to them. We overlaid the resident and visitor maps separately to produce two composites that depict each group's evaluative image of the city. The composite maps show the places that respondents said they liked visually and the places that they said they disliked, as well as the degree of consensus on these evaluations. The composite maps also point to physical properties associated with the evaluations.

We found three important aspects of the evaluative image: *identity, location,* and *likability.* The maps show the salient elements (identity), where they are (location), and the qualities associated with a strong evaluative response among the public (likability). This information has value for the design, planning, and management of city appearance because it provides specific information on visual quality for specific places throughout the city.

Figure 3.1. Aerial View of Knoxville from the Southwest. This view shows the CBD Center left, the Hyatt Hotel and Coliseum-Auditorium just beyond, and the University and Neyland Stadium lower right. Copyright Gary Heatherly; used with permission.

Knoxville

Knoxville is in central Tennessee along the Tennessee River and at the intersection of two major interstate highways: I-40 and I-75. In 1798, Charles McClung plotted the city into 16 blocks with 64 lots. According to census figures, by 1980, Knoxville's urbanized area had 384,708 people, a 50% increase from 1970, and it continued to increase by another 50% to 585,960 people by 1990 (U.S. Department of Commerce, 1983).

The city has an old commercial core. This 240-acre central business district (CBD) is like an island (Figure 3.1). Creeks and river valleys surround the CBD on a plateau. Motorists can reach it only by bridge. A major road (Gay Street) runs north and south through the center of the CBD. It has highways on three sides (I-158 ringing it along the south and east, I-40 to the north) and an abandoned rail yard on the fourth side. (After this research, the rail yard became the site of the International Exposition.) Farther west is Henley/Broadway-441. Nine older inner-city

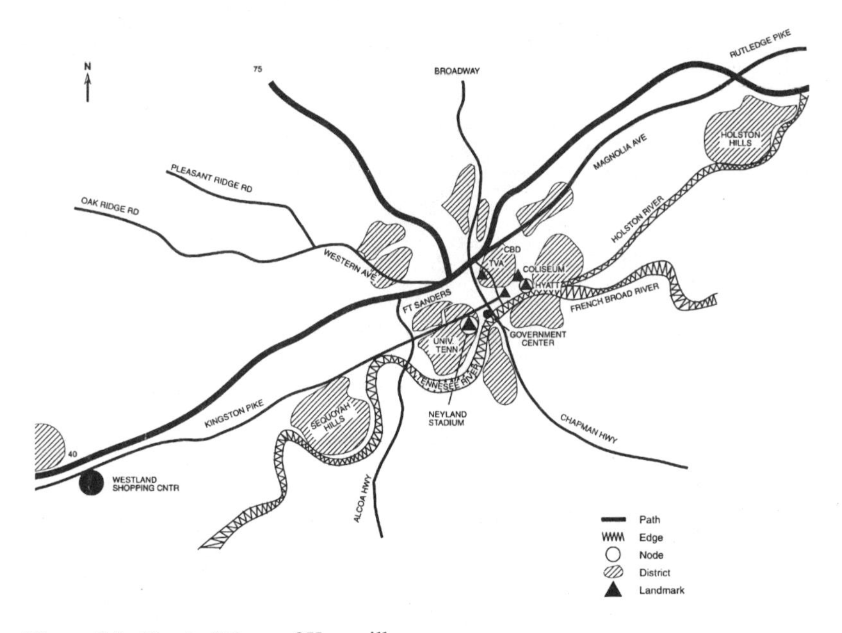

Figure 3.2. Physical Form of Knoxville
Drawing by David Miller.

neighborhoods ring the CBD (Fort Sanders to the west, Morningside Heights to the east, Sevier to the south, and Beaumont, Mechanicsville, North Knoxville, and Fourth-and-Gill to the north). In addition, the University of Tennessee and its landmark Neyland Stadium sit west of downtown between a major road (Cumberland/ Kingston Pike) and the Tennessee River. Knoxville had one major visible new building downtown (a bank) as well as urban renewal developments on grassy hills east of downtown—the distinctive and visible Hyatt Hotel and Coliseum-Auditorium. Farther out are two older suburban areas (Sequoyah Hills to the west and Holston Hills to the east) and still farther out are newer suburban developments such as Westwood and Cedar Bluff. One highway (Alcoa Highway, running southwest) and five major roads (Magnolia Avenue to the east, Broadway to the north, Western Avenue to the northwest, Cumberland Avenue/Kingston Pike to the west, and Chapman Highway to the south) fan out radially from the CBD. Development also fans out radially from the CBD and the Tennessee River to the hills,

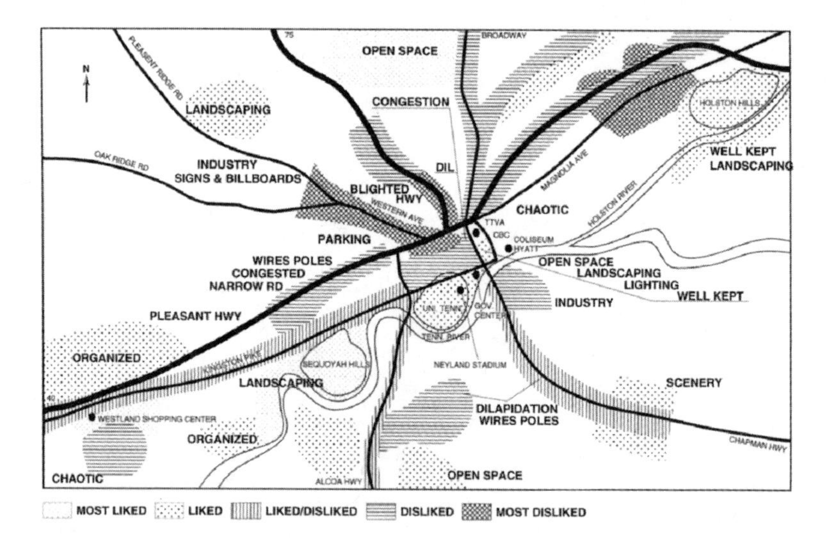

Figure 3.3. Evaluative Map of Knoxville from Verbal Descriptions by Residents
Drawing by David Miller.

which afford views of the Smoky Mountains 35 miles to the south. A heavily used national park, the Smokies represent a valuable resource (in

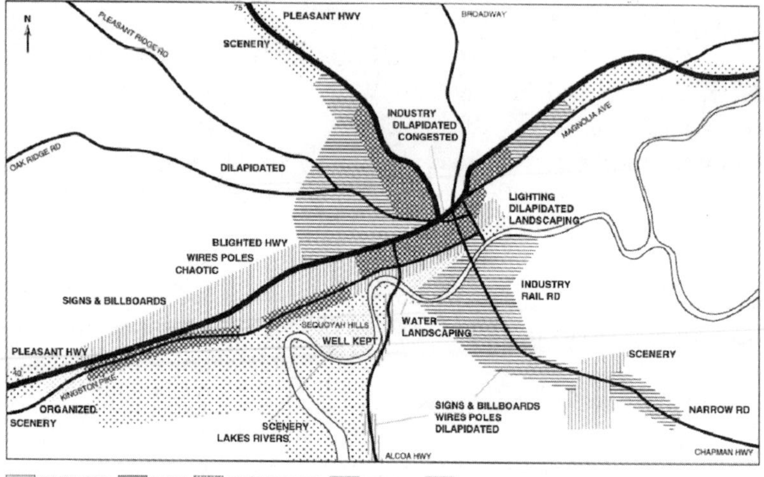

Figure 3.4. Evaluative Map of Knoxville from Verbal Descriptions by Visitors
Drawing by David Miller.

scenery, outdoor activities, and tourism) to the city. Knoxville provided one third of the public funding to purchase the land for the original park.

A physical inventory of the city reveals prominent elements that seem to fit Lynch's (1960) classification as districts, paths, edges, landmarks, and nodes (see Figure 3.2). The elements that might stand out in people's image as districts include the CBD, neighborhoods ringing the CBD, the University of Tennessee, the two older suburban neighborhoods, and the newer suburbs. Each of these areas appears to have a recognizable character. The elements that might stand out in people's images as paths include I-40, I-70, Gay Way, the major roads fanning out from the CBD, and the Tennessee River. They all represent high-use channels of movement. The Tennessee River becomes an intensely used path periodically when people travel by boat to and from football games. Some of the paths may also stand out as edges. They include I-40, I-70, the Tennessee River, the rail yard, and the highways around the CBD. All these linear elements create boundaries. Elements that might stand out in people's images as landmarks include the Smoky Mountains, the bank tower, Neyland Stadium (Figure 3.10), and the Hyatt Hotel and Coliseum Complex (Figure 3.15). They may stand out for their identity and as external points of reference. For example, people can see the mountains from many points throughout the area; at the time of the study, the bank tower, built by a banker who had become governor, stood as the tallest and newest building downtown. Elements that might stand as nodes include Westown Mall (Figure 3.16), the Hyatt Hotel, the coliseum, Neyland Stadium, and the airport (Figure 3.17). Each attracts intensive use.

The maps derived from the Knoxville interviews depict the evaluative images of the city (Figures 3.3 and 3.4). They display the well-known (imageable) places that many observers reported either liking or disliking. As expected, the resident maps appear to cover more places and make finer discrimination between places than the visitor maps. Though the differences may relate to the different interview methods, we would expect the responses to the map (visitor interviews) to provide more detailed information than simple verbal recall over the phone. Both groups agreed on their evaluations. The maps show that both residents and visitors dislike the appearance of chaotic and rundown districts and paths—that is, the CBD, the streets feeding it, and the inner-city neighborhoods. They like the look of new, spacious

TABLE 3.1 Knoxville: Prominent Disliked Elements and Reasons

Disliked Element	*Reported Reasons*
CBD and Gay St. (Figure 3.5)	Parking lots Poles, wires, and signs Industry Dilapidation Abandoned rail yard Dark canopied street Lack of coherent styles
Inner-City Neighborhoods—Fort Sanders, Morningside Heights, Sevier, Beaumont, Mechanicsville, North Knoxville, and Fourth-and-Gill (Figure 3.6)	Dilapidation Dirtiness
Highways—I-40, I-75, Alcoa Highway (Figure 3.7)	Billboards Signs Industry
Major Streets—Magnolia Ave., Broadway, Western, Cumberland/ Kingston Pike, Chapman (Figure 3.8)	Poles and wires Industry Poor upkeep Congestion Chaotic outdoor advertising

Figure 3.5. Knoxville CBD, an Abandoned Rail Yard, Disliked for Dilapidation, Parking Lots, Poles, Wires, Signs, Industry, a Dark Canopied Street, and Lack of Coherent Styles
Photographs by David Bokenkamp.

Figure 3.6. Low-Income Inner-City Neighborhoods: Disliked for Dilapidation and Dirtiness
Photographs by David Bokenkamp.

landmark centers, vegetated districts, and paths with views of the river and the Smokies.

Disliked Areas

Table 3.1 culls from the maps the prominent disliked elements and the reasons respondents reported disliking them. Respondents gave negative evaluations to the CBD, several highways and roads, and several inner-city neighborhoods (Figure 3.5 - 3.8). They said they disliked these areas because of such features as poles, wires, billboards, and signs, parking lots, dilapidation, poor upkeep, dirtiness, congestion, a dark canopied street, and lack of coherent styles. (The city has constructed a bypass to alleviate the congestion, but at the time of the study, the intersection of I-40 and I-75 had so much congestion that truckers and others referred to it as *malfunction junction*.)

Figure 3.7. Highways (I-40 and I-70): Disliked for Industry, Signs, and Billboards
Photograph by David Bokenkamp.

Figure 3.8. Major streets (Kingston Pike): Disliked for Chaotic Outdoor Advertising, Poles and Wires, Industry, Poor Upkeep, and Congestion
Photographs by David Bokenkamp.

Liked Areas

Table 3.2 lists the prominent liked elements (Figures 3.9, 3.10, 3.11, 3.12, 3.13, 3.14, 3.15, 3.16, 3.17) and the reasons respondents reported for liking them. Respondents liked the appearance of several districts—the University of Tennessee and the two older suburban neighborhoods, Sequoyah Hills and Holston Hills (Figures 3.9, 3.10, 3.11). Perhaps because of the remoteness of Holston Hills and its location on the slower-growing,

Figure 3.9. The University of Tennessee Campus: Liked for Its Well-Kept Historical Buildings, Abundant Greenery, Views of the Smoky Mountains, and the River
Photograph by David Bokenkamp.

Figure 3.10. Neyland Stadium at the University of Tennessee Campus: Its Large Size, Intense Use as a Football Stadium, and Symbolic Meaning May Contribute to Its Identity as a Landmark
Photograph by David Bokenkamp.

TABLE 3.2 Knoxville: Prominent Liked Elements and Reasons

Liked Element	*Reported Reasons*
University of Tennessee (Figures 3.9, 3.10)	Profuse vegetation Views (river, mountains) Historical buildings
Sequoyah Hills and Holston Hills (Figure 3.11)	Landscaping Well-kept Cleanliness Views (mountains, river)
Westwood and Cedar Bluff (Figure 3.12)	Organized
Green Roads Exurban highways (Figure 3.13) Alcoa, I-40, I-75, Chapman In-town highways (Figure 3.14) Alcoa by River, Cumberland	Landscaping Openness Views (mountains, nature, river)
Hyatt Hotel and Coliseum-Auditorium (Figure 3.15)	Greenery Design Open Space
Westown Mall (Figure 3.16)	Clean Organized
Airport (Figure 3.17)	Clean Modern Views (countryside, mountains)

Figure 3.11. Sequoyah Hills: Liked for Its Landscaping, Views of the River and Mountains, Cleanliness, and Good Upkeep

Located west of the University of Tennessee campus, this older (higher-status) neighborhood has large homes, access to the river, and views of the Smokies. The Dogwood Trails attract thousands of people in the spring for the annual Dogwood Festival. Photograph by David Bokenkamp.

Figure 3.12. Westwood/Cedar Bluff: Liked Newer Suburb for Its Organization
These newer suburbs that have sprung up along I-40 west have low-density single-family homes on large lots. These also stand out as exits on I-40.
Photographs by David Bokenkamp.

lower-income side of town (Figure 3.12), visitors mentioned it less often. Respondents also liked newer suburbs (Westwood/Cedar Bluff) on the faster growing side of town. They reported that they liked several green

Figure 3.13. Exurban Streets: Liked for Landscaping, Openness, and Views of the Countryside (Trees and Fields) and Mountains
Photograph by David Bokenkamp.

Figure 3.14. Cumberland Avenue: Liked for Abundant Vegetation, It Runs by the High-Status, Well-Liked Sequoyah Hills
Photograph by David Bokenkamp.

Figure 3.15. Hyatt Hotel (right) and Coliseum-Auditorium (left): Liked for Their Greenery, Open Space, and Design

Their visibility atop a grassy hill across from downtown and the distinctive typewriter shape of the Hyatt may contribute to their identity as landmarks.
Photograph by David Bokenkamp.

Figure 3.16. Westown Mall: Liked for Its Clean, Orderly Appearance, It May Benefit from Its Contrast with the Adjacent Chaotic Commercial Development (Figures 3.7 and 3.8)
Photograph by David Bokenkamp.

highways: exurban portions of Alcoa Highway, I-40, I-75, and Chapman Highway, and, closer to downtown, a riverfront section of Alcoa Highway and a vegetated section of Cumberland (Figures 3.13 and 3.14). Finally, they said they liked the appearance of the three newer developments—the Hyatt Hotel and Coliseum-Auditorium, the airport, and the Westown Shopping Mall (Figures 3.15, 3.16, 3.17). They said they liked these areas for such features as vegetation, countryside, landscaping, upkeep, cleanliness, openness, views (to mountains, rivers, nature), organization, design, and historical buildings.

Figure 3.17. Knoxville Airport, Located in the Countryside: Liked for Its Clean, Modern Appearance, and Views of the Countryside and the Smoky Mountains
Photograph by David Bokenkamp.

Chattanooga

Chattanooga is located in southeast Tennessee at the Georgia border. Originally the town of Ross Landing, on the Moccasin Bend of the Tennessee River, the city was laid out on 240 acres and incorporated as Chattanooga in 1838. According to census figures, by 1980, the Chattanooga urbanized area had increased more than 30% since 1970 to 301,515 residents, and it increased an additional 40% to 424,347 residents in 1990 (U.S. Department of Commerce, 1983).

Unlike downtown Knoxville, situated above its surroundings, downtown Chattanooga lies in a valley surrounded by mountains. These include Lookout Mountain (2,391 feet) to the southwest, Missionary Ridge (1,100 feet) to the east, and Signal Mountain (Shoen, 2,080 feet) to the northwest (Figure 3.18).

Lookout Mountain and Missionary Ridge dominate the city. The city's name comes from a Native American word for Lookout Mountain ("rock rising to a point"). The mountains have historical significance as Civil War battlefields, and they attract heavy tourist use. Lookout Mountain's tourist attractions include the Incline Railway, mountain caves, the Mountain Bridge, and a lookout from which observers can see seven states. Missionary Ridge (now part of the Chickamauga and Chattanooga National Military Parks) has several military monuments.

Figure 3.18. Aerial view of Chattanooga from the Southwest Showing I-24 in the Right Foreground and the CBD in the center.
Courtesy of Tennessee Valley Authority.

The Chattanooga Choo Choo, a restored Victorian railway station, reflects Chattanooga's historic importance as a major rail center. With a hotel, working train, dining space for 1,300 people (including an 85-foot dome), shops, and entertainment, the Choo Choo has attracted intense use.

The Chattanooga CBD has a mix of modern high-rise buildings and turn-of-the century ornamental buildings. Many buildings look dilapidated. Moving south of the CBD, one encounters railroad tracks, industry, and housing. Inner-city housing in the 9th Street area and neighborhoods south of the CBD, including Alton Park, St. Elmo, and Piney Woods, have bungalows, duplexes, tenement houses, and shacks. Many residents abandoned these rundown neighborhoods to move to the suburbs. To the north, North Chattanooga and Hixon have a mix of old homes, apartments, and offices. Other well-known neighborhoods include Missionary Ridge, Brainerd, East Ridge, Rossville, and Redbank. Major roads include I-24 (west), I-27 (north-south), I-75 (north-south), Cummings Highway, Broad Street (through downtown), Ochis Highway (south), and McClue and Bailey Avenues through downtown.

The physical inventory of Chattanooga (Figure 3.19) reveals elements likely to stand out to observers as districts (the CBD and various neighborhoods), paths (various highways and major roads), edges (the river, highways, and mountains), landmarks (the Chattanooga Choo Choo), and nodes (the Choo Choo, CBD, Rock City, the Lookout Mountain tourist attractions).

Following similar procedures to those used to study Knoxville, we interviewed 60 residents and 60 visitors in Chattanooga. We asked them to identify the areas they liked visually and the areas they disliked visually and to give the reasons for their responses. (Chapter 5, "Evaluating the Method" describes and evaluates the method in detail.) As with Knoxville, the evaluative maps of Chattanooga reveal agreement on likability, suggesting ways to improve city appearance.

Disliked Areas

As with Knoxville, the evaluative maps of Chattanooga (Figures 3.20 and 3.21) show that residents cover more places and make finer discrimination between places than visitors. Both groups show agreement on their dislikes and likes. Table 3.3 lists the prominent disliked elements and the

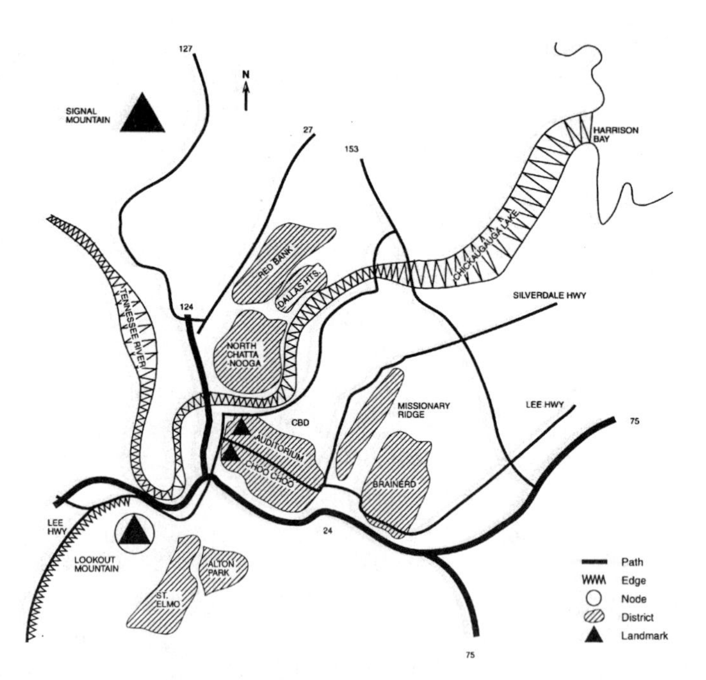

Figure 3.19. Physical Form of Chattanooga
Drawing by David Miller.

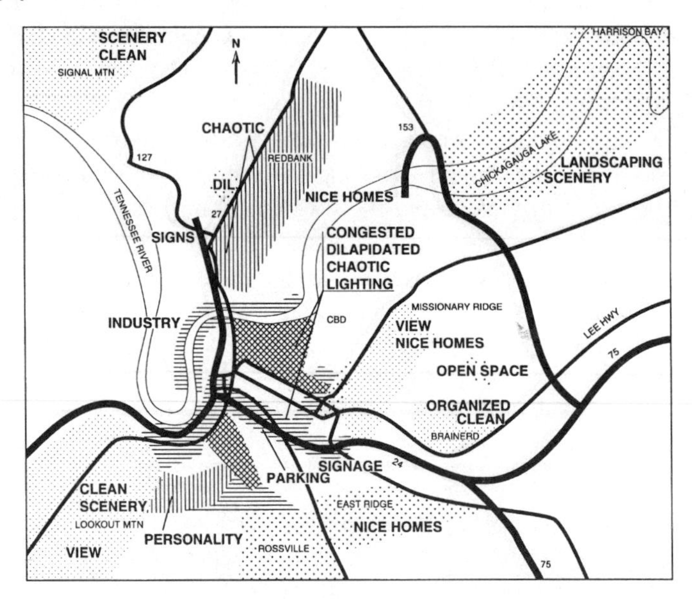

Figure 3.20. Evaluative Map of Chattanooga from Verbal Descriptions by Residents
Drawing by David Miller.

TABLE 3.3 Chattanooga: Prominent Disliked Elements and Reasons

Disliked Element	*Reported Reasons*
CBD (Figure 3.22)	Parking lots Dilapidation Pollution Congestion Chaotic signs Lighting
Inner-City Neighborhoods (Figure 3.23)	Dilapidation Disorder
Interstate 24 and South Broad (Figure 3.24)	Signs Industry Crowding
Commercial Area in North Chattanooga (Figure 3.25)	Signs Parking lots Dilapidation Litter Pollution Congestion, crowding Disorder, confusion Lighting

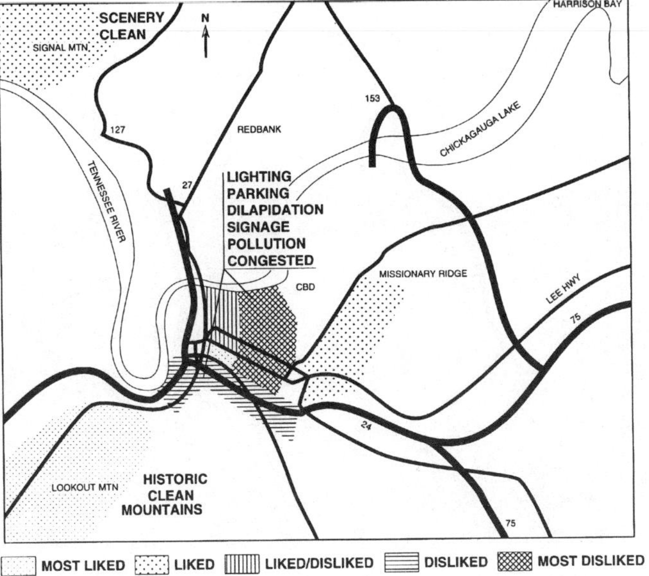

Figure 3.21. Evaluative Map of Chattanooga from Verbal Descriptions by Visitors
Drawing by David Miller.

Figure 3.22. Chattanooga CBD: Disliked for Parking Lots, Lighting, Dilapidation, Pollution, Congestion, and Chaotic Signs

Figure 3.23. Chattanooga Inner-City Neighborhood: Disliked for Dilapidation and Disorder
Photograph by David Bokenkamp.

reasons that respondents gave for disliking them. Respondents disliked rundown districts near downtown and chaotic paths. They disliked several districts including the CBD, the Ninth Street area, and several neighborhoods south of the CBD (Figures 3.22 and 3.23). They also disliked several paths: an industrial part of I-24, South Broad, and the crowded, commercial sections of North Chattanooga/Hixon (Figures 3.24 and 3.25). Respondents

Figure 3.24. I-24 and South Broad: Disliked for Industry, Crowding, Chaotic Signs
Photographs by David Bokenkamp.

Figure 3.25. Commercial Area in North Chattanooga: Disliked for Parking Lots, Lighting, Dilapidation, Pollution, Litter, Disorder and Confusion, Crowding, Congestion, and Signs
Photograph by David Bokenkamp.

often cited parking lots, signs, dilapidation, pollution, congestion, disorder (confusion), and lighting as reasons for these negative evaluations.

Liked Areas

Table 3.4 lists the prominent liked elements and the reasons respondents reported liking them.

Respondents liked some residential districts (Figures 3.26 and 3.27). They also liked two sets of landmark/nodes—the mountains (Signal Mountain, Lookout Mountain, and Missionary Ridge) and the historical

TABLE 3.4 Chattanooga: Prominent Liked Elements and Reasons

Liked Element	*Reported Reasons*
Neighborhoods	
Missionary Ridge, Brainerd, East Ridge, Rossville (Figure 3.26)	Clean
	Views
	Organized
	Nice homes
Redbank, North Chattanooga (Figure 3.27)	New single-family homes
Mountains	
Signal Mountain, Lookout Mountain, Missionary Ridge (Figure 3.28)	Scenery
	Historical significance
	Clean
Chattanooga Choo Choo (Figure 3.29)	Significance

Chattanooga Choo Choo (Figures 3.28 and 3.29). They cited landscaping, cleanliness, openness, views, organization, and historical significance as reasons for their preferences.

Figure 3.26. Rossville: One of Four Neighborhood Liked for Views, Cleanliness, Organization, and Nice Homes
Photograph by David Bokenkamp.

Figure 3.27. North Chattanooga near Hixon: One of Two Neighborhoods Liked for Newness of Homes
Photograph by David Bokenkamp.

Figure 3.28. Views Toward Mountains: Liked for Their Scenery, Historical Significance, and Cleanliness
Photograph by David Bokenkamp.

Figure 3.29. The Chattanooga Choo Choo: Liked for Its Historical Significance
Photograph by David Bokenkamp.

Relevance of the Evaluative Responses

The evaluative maps in each city reveal agreement on likability and suggest ways to improve the appearance of each city. Not all the likely imageable elements appear on the maps, probably because people did not feel strongly about them favorably or unfavorably. Given the high visibility of the places that do appear on the maps, changes in their appearance could have a strong effect on the public image of the cityscape. But what should be changed, what should be reinforced? The evaluative maps show for each place the reasons given for the evaluations. Taking any given site, these reasons point to specific directions for improvement. For example, the responses to the disliked portions of I-40 and I-75 in Knoxville suggest that the city could improve these areas by reducing the prominence of chaotic signs, billboards, poles, and wires.

A broader set of directions emerged from tallying the reasons across each city. These tallies reveal negative evaluations associated with chaotic commercial development, signs and billboards, dirtiness, weeds, narrow bridges, dilapidation, poles, wires, and industry. They reveal favorable evaluations associated with landscaping, countryside, scenery,

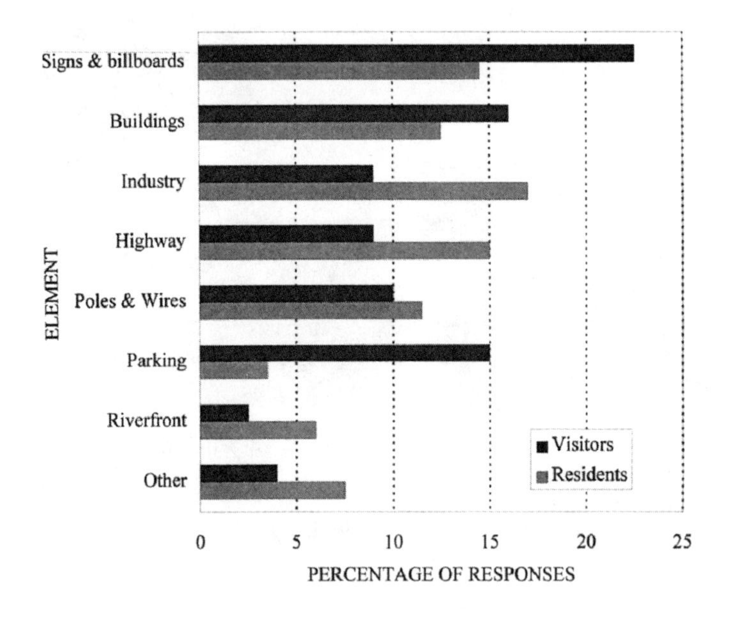

Figure 3.30. Knoxville: "Select From This List of Elements the One Most in Need of Visual Improvement"

NOTE: a. Resident versus visitor responses ($\chi^2 = 22.1$, 1df, $p < 0.01$); comparison of elements by residents, ($\chi^2 = 22.9$, 7df, $p < 0.01$), visitors ($\chi^2 = 30.84$, 7df, $p < .01$), and all respondents ($\chi^2 = 33.8$, 7df, $p < 0.01$)
b. 13% of residents and 12% of visitors gave no response.
c. "Other" includes separate references to railways, gas stations, and buses.

vistas, new buildings, topography, and organization. To supplement the open-ended evaluative responses, we asked people in Knoxville to choose from a list of features the one most in need of improvement. Signs/billboards, buildings, and industry registered the strongest responses (Figure 3.30).

The results parallel those shown on the evaluative maps. Residents complained most often about industry, followed by highways, signs and billboards, buildings, poles and wires, the riverfront, and parking. Visitors complained most often about signs and billboards, followed by buildings, parking, poles and wires, industry, highways, and the riverfront.

In Chattanooga, we replaced the list with an open-ended question: "What one thing should be done to improve the city's appearance?"

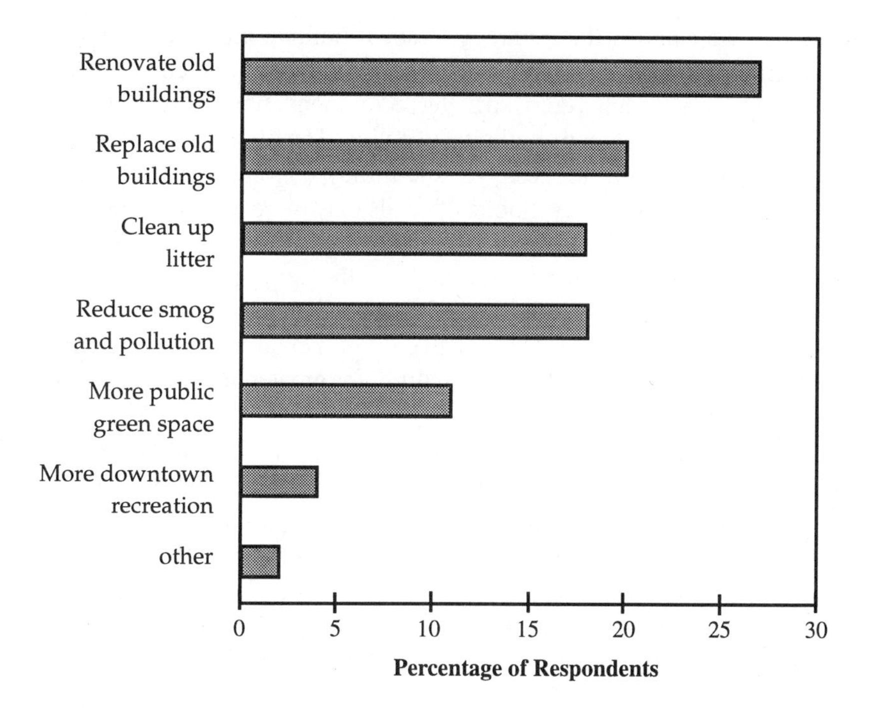

Figure 3.31. Chattanooga: "What One Thing Should Be Done to Improve the City's Appearance?" (n = 120)

Respondents most frequently mentioned renovating or replacing old buildings, followed by cleaning up litter, reducing smog and pollution, increasing green public space, and increasing downtown recreation areas (Figure 3.31)

In sum, the results suggest that each city could enhance its evaluative image by increasing the visibility of vegetation, landscaping, mountains, rivers, and historical elements and by decreasing the visibility of outdoor advertising, poles and wires, parking lots, unsightly industry, and dilapidation.

Many designers might have come up with similar suggestions intuitively. Do communities need all the research on the evaluative image to find that people like natural areas and dislike chaotic ones? The agreement of the findings with intuition does not make the method or approach irrelevant. Sometimes social science research confirms the expected, and sometimes it yields counterintuitive findings. Only by conducting the study can we can test the accuracy of such intuitive hunches. Furthermore, although designers may have come up with similar recommendations, the survey findings demystify the recommendations by making the process for arriving at them explicit and public. The findings also depend on a more democratic process. As such, they can become useful for implementing public policy. Public officials may more willingly support recommendations based on public opinion than those based on an expert's appraisal because the survey captures the opinions of constituencies. In addition, surveys as a form of "consumer research in the public sector," (Chadwin, 1975, p. 47) fit with environmental legislation (Zube, 1980) and Supreme Court decisions calling for greater accountability of state decisions (Stamps, 1996).

Evaluative maps may encourage cosmetic treatments for deeper structural problems. This risk has more to do with the use of the maps than the maps themselves. I expect users of these maps to consider the evaluative image and meaning along with other concerns—such as social structure, land use, and transportation—that define a plan. Visual quality alone may not justify environmental change, but we should not overlook it as irrelevant. People see visual meanings as important. The maps reveal public agreement on likability.

With the evaluative images of two cities in hand, we can compare the components of likability in the two cities. Is there a shared structure of preferences and dislikes in each city? How do city structure and experience influence the evaluative image? The next chapter considers these questions.

4

The Elements of Urban Likability

Evaluative maps reflect the public evaluative images of the city. These public images consist of the overlapping individual images of many people. As the resident and visitor maps discussed in Chapter 3, "Two Cities" suggest, distinct public images are held by large numbers of people in different groups. Meanings in the evaluative image are predictable in the aggregate. Though subculture aggregates may differ from one another (Royse, 1969), visual preferences are highly stable. Just as city imageability or legibility helps people operate within a city (Lynch, 1960), evaluative meanings may affect people's movements in a city. They may influence the choice of neighborhood, places to shop, places for recreation, and travel routes. Of course, appearance may not shape behavior in all situations. An ugly workplace may not keep people from going to work. Other factors such as convenience, price, and familiarity may constrain the choice. Given a real choice, however, people would probably rather go to attractive places and avoid unattractive ones. Good appearance should also relate to the delight people take in a place, how well they remember it, and whether they come back to it for itself.

Before examining the map in more detail, we should consider the distinction between prediction and cause. The establishment of cause requires a covariance between cause and effect, the temporal precedence of cause, the ability to rule out of rival hypotheses, and the manipulation of the assumed cause (Cook & Campbell, 1979). Researchers achieve this through controlled experiments. (Chapter 5, "Evaluating the Method," and the appendix discuss some techniques for doing this for the evaluative image.) One does not need to establish causality to use a finding to forecast likely outcomes of design manipulations. Forecasting is less concerned about why the predictor works. If manipulating specified variables results in improvements in the evaluative image, the manipulation has achieved the desired effect, even if one has not identified the cause. The Knoxville and Chattanooga studies did not involve controlled experiments. They used nonexperimental designs that did not allow the establishment of cause. Still, planners can use the results—patterns of associations found between environmental features and preference—to forecast likely outcomes of design manipulations.

In making predictions from the results, one assumes that the preferences represent a "stable predisposition to respond" (Cook & Campbell, 1979, p. 64). The Knoxville and Chattanooga studies obtained one set of responses, one measure of the effect (liking), and several related methods. They used verbal recall, recall responses on maps, verbal reasons, and ratings of lists of elements. From this evidence alone, we cannot know if the likable features have stability across other kinds of responses, measures, and methods and more generally to citizens' daily experiences. By looking at other research using different methods, populations, and contexts, however, we can get a better sense of the stability of the findings. Toward that end, this chapter discusses the predictions for likability.

A glance at the maps of Knoxville and Chattanooga reveals a pattern of responses that agrees with my earlier assessment of the visual quality of American cities. Although the maps show some likable areas, the prevalence of darker shades in the imageable areas indicates a neutral-to-negative image. This, in turn, highlights the need to improve the evaluative image.

What kinds of physical features do people remember as likable?[1] The evaluative maps in each city include the five elements cited by Lynch (1960)—paths, edges, districts, nodes, and landmarks—with some liked, others disliked. Respondents' dislikes tended to center on districts and paths; their likes tended to center on districts, paths, and landmarks. Imageability is not enough. The maps have a second component beyond imageability: emotional meaning (likability). For a favorable image, features must stand out as both memorable and likable. The four maps reveal similarities in the environmental features associated with these two components; this consensus suggests some directions for urban design decisions. First, let's consider imageability.

Distinctiveness, Visibility, and Use/Symbolic Significance

As with cognitive maps (deJonge, 1962; Gulick, 1963), the evaluative maps of Knoxville and Chattanooga contain broad aspects of the city, such as roads and districts. Researchers have found three factors that affect the imageability of elements: distinctiveness of form, visibility, and use/symbolic significance (Appleyard, 1976; Evans, Smith, & Pezdek, 1982). Although we did not directly measure these factors, the maps show associations with visibility, significance, and, to a lesser extent, distinctiveness. The maps also include both built and natural elements. The central business districts (CBDs) and the commercial strips in both cities, the hotel and coliseum in Knoxville, and the mountains in Chattanooga have high visibility and use significance. The Chattanooga Choo Choo and Knoxville's Westown Mall, Sequoyah Hills, and Holston Hills have use significance. The Chattanooga Choo Choo, the old railroad center of Chattanooga, has been remodeled into a hotel and Victorian-style restaurant—a large gathering place. In Knoxville, the Westown Mall is a major shopping mall in the city. The Dogwood Trails attract thousands of visitors to Sequoyah Hills and Holston Hills every spring to view the dogwood trees and flowers in bloom. In Chattanooga, the Choo Choo and mountains also have historical significance, the latter for various Civil War sites. The Knoxville Hyatt Hotel and Coliseum-Auditorium and the Chattanooga Choo Choo also have distinctiveness: Viewed from a hill across from the CBD, the hotel and coliseum have unique forms; the Choo Choo stands out as a large Victorian station. Notice that some elements (such as the Choo Choo, the mall, and the hotel and coliseum) have a convergence of distinctiveness, visibility, and use significance. Others (such as the CBD and Sequoyah Hills) do not. In sum, it appears that the evaluative image relates to distinctiveness of form, visibility, and use significance, or the convergence of these factors.

Likable Features

Though the structures of the maps differ, residents and visitors in each city gave similar reasons for their evaluations. I have classified these reasons into five kinds of environmental attributes—naturalness, upkeep/civilities, openness, historical significance, and order. The liked areas tend to have these attributes and the disliked areas tend to have their opposites—obtrusive, humanmade uses; dilapidation; restriction; a lack of any historical significance; and disorder. (We did not find respondents reporting a dislike for new buildings.) The likable attributes may also have a more general application because they appear in theory and research on environmental preferences, described later in this chapter. I define these elements as follows:

1. *Naturalness* refers to the presence of vegetation, water, or mountains. Respondents reported that they liked places for landscaping, countryside, rivers, lakes, water, and mountains. They reported dislikes for built areas of high contrast, referring to the appearance of commercial strips, industry, poles, wires, and signs.
2. *Upkeep/civilities* refers to the maintenance of areas. Respondents reported that they like places for their cleanliness, maintenance, and new homes. They reported disliking places for their dilapidation, dirtiness, weeds, and lack of upkeep. Some researchers refer to these disliked features as *physical incivilities* because the features serve as cues to social disorder (Taylor, 1989).
3. *Openness* refers to the vista. People often reported liking places for the presence of open space and scenery. They often reported disliking places for their restriction, crowding, congestion, and narrow roads.
4. *Historical significance* refers to places perceived as having historical significance. Places may have authentic historical significance or look historical to the observers. In either case, such places evoke favorable response. People often said they liked a place for its historical appearance or associations.
5. *Order* refers to the degree to which respondents feel an area looks organized. Respondents reported that they liked areas for their visual order, referring to order, cohesiveness, and compatibility. They said they did not like areas with disorder, referring negatively to chaos and the lack of uniform style. Compatibility or the degree to which features in a scene fit with one another also provides order, whereas incompatible elements or nonuniform styles lessen order.

Figure 4.1. Natural Features Tend to Evoke Favorable Responses
Photograph by Jack L. Nasar.

How do the five features found in Knoxville and Chattanooga hold up in relation to other research? Are the features prominent in perception and preference?

Naturalness

In Knoxville and Chattanooga, people responded strongly to the naturalness of places (Figure 4.1). They like natural areas, and they dislike areas with intense land uses such as industry. Other research confirms that people notice differences in the naturalness of places and prefer naturalness. The natural-built dimension has repeatedly emerged as perhaps the most prominent dimension of human response to the environment (Herzog, Kaplan, & Kaplan, 1976, 1982; Nasar, 1988b, 1989a, 1994). Humans see natural (vegetation) and built scenes as two distinct and prominent content categories (Kaplan & Kaplan, 1989). When they sort photographs of different places, a distinction between natural and built content often underlies the sorting (Ward & Russell, 1981; Wohlwill, 1983). Research has also found a related variable—intensity of use—prominent in people's perception (Hanyu, 1995; Herzog et al., 1976, 1982). The distinction between natural and built content may start early: Young children make a distinction between living and inanimate things (Piaget, 1962).

How do we define a natural as opposed to a built setting? Naturalness typically involves the user's perception of an area as natural or the predominance of natural over built elements (Wohlwill, 1983). Thus, some "natural" settings depend on human intervention, others have built elements, and others exist within built contexts.

The pattern of response found in Knoxville and Chattanooga agrees with a substantial body of experimental and nonexperimental research that finds human preferences for natural over built features or perceived human intervention (Kaplan & Kaplan, 1989; Nasar, 1994; Ulrich, 1983; Wohlwill, 1976). Lynch (1960) found that people noted vegetation or water "with care and pleasure" (p. 44). Although mapping data refer to recall, many other studies refer to perceptions. Studies show naturalness as a powerful predictor of preference (Nasar, 1988c). Moreover, they show that adding vegetation to scenes increases preferences (Thayer & Atwood, 1978) and that people prefer natural scenes to scenes perceived as having human intervention (Kaplan, Kaplan, & Wendt, 1972; Wohlwill, 1974, 1983). Research also confirms a dislike for intense development, such as high-rise buildings and commercial and industrial uses (Herzog et al., 1976, 1982; Wohlwill, 1982). Two studies show an effect on patterns of movement in a city: Lynch (1960) notes that several people reported "daily detours which lengthened their trip to work but allowed them to pass some particular planting, park or body of water" (p. 44). A controlled study confirms that people drive out of their way to use a parkway rather than a more direct but less natural expressway (Ulrich, 1973).

Vegetation may also strengthen the imageability of elements. Lynch (1960) reports that a great deal of planting along a path reinforces its image; he suggests that paths along water or along parks tend to be more memorable (p. 51). A study of features associated with building recall and location memory confirms Lynch's speculation: The presence of nature around the building contributed to building imageability as a landmark (Evans et al., 1982).

Beyond that, a body of research has begun to suggest a calming and restorative value of nature (Kaplan, 1995; Kaplan & Kaplan, 1989; Kaplan & Talbot, 1983). Patients with a window view of deciduous trees have faster postoperative recovery, fewer negative evaluations by nurses, and fewer doses of narcotic painkillers than patients with a view of a brick wall (Ulrich, 1983). Exposure to nature-based activities has positive effects on cancer patients (Kaplan, 1995). Individuals viewing videotapes of nature show more rapid psychophysiological recovery than those viewing tapes of urban scenes (Ulrich et al., 1991). People walking through a natural area show higher levels of restoration from stress than people walking through a built one (Hartig, Mang, & Evans, 1991). Outdoor challenge programs and wilderness vacations also show positive effects (Kaplan & Kaplan, 1989; Ulrich et al., 1991). The effects found may result from a more complex process. For example, the positive emotional experience associated with nature may mediate the restorative effect, or the character of the setting may affect which feature becomes restorative. Still, it appears that we can predict restorative effects associated with nature.

The consistent preference for nature may relate to either the *content* or the *form* of natural elements. Naturalness and vegetation may evoke favorable associations and connotative meanings associated with content. If you see a park and associate it with lovers, recreation, and leisure, the associations may color your evaluative response. One study found that respondents associate the presence of vegetation with status, and groups differ in the meanings they associate with manicured versus natural landscape (Royse, 1969). The preference for nature may simply relate to its form. Unlike built elements, natural features may have more gradual changes; irregular and curvilinear lines; continuous gradation of shape and color; irregular, rougher textures; smoother, less intense, less predictable irregularities; movement; and sound (Wohlwill, 1979, 1983). They may have repetitions across scales, such as fractals, not present in the built environment. By buffering the more chaotic built elements, nature may also add order. According to one hypothesis, we evolved to prefer nature because of the millions of years our ancestors spent in the African savanna (Ulrich, 1983). Whatever the cause, cities can improve their evaluative image by adding natural elements (such as trees, water, and mountains) and providing views to nature.

Upkeep/Civilities

Another set of prominent features on the evaluative maps has to do with the care and upkeep of the environment (Figure 4.2). These features also have emerged as prominent in human perception of the environment (Herzog et al., 1976. 1982; Nasar, 1988c, 1989a, 1994). The findings in

Figure 4.2. People Tend to Like Well-Maintained Places
Photograph by David Bokenkamp.

Knoxville and Chattanooga that people recall areas as disliked because of dilapidation, poles, wires, signs, and vehicles agree with many findings on environmental preference (Cooper, 1972; Marans, 1976). For example, Lynch (1960) reports references to "old," "dirty," and "drab" for Jersey City (p. 50). A study relating evaluative responses (preference, interest, and safety) to judged physical features of residential scenes found upkeep as a primary predictor of evaluative response (Nasar, 1988c). A study that manipulated features in photographs one variable at a time identified exterior maintenance as having a strong influence on social meanings inferred (Royse, 1969). Another study that obtained evaluative responses to photographs of scenes systematically altered to remove utility poles, overhead wires, billboards, and signs separately found that the removal of these features improved the evaluative image of roadside scenes (Winkel, Malek, & Thiel, 1969). A controlled study of retail signs found that reductions in sign size and contrast enhanced the evaluative image of a retail scene (Nasar, 1987). Studies on fear of crime and crime show that increased fear and actual crime relate to physical incivilities such as dilapidation or a milieu showing an absence of care (Newman, 1972; Taylor, 1989). Studies of traffic show that lower levels of traffic elicit favorable changes in the evaluative image and quality of life for residential streets and neighborhoods (Appleyard, 1981; Craik, 1983; Lansing, Marans, & Zehner, 1970).

The findings for upkeep/civilities agree with findings of status associated with these features (Duncan, 1973; Royse, 1969) and with findings of preference for order and natural materials over disorder and built materials. The disliked features—such as incivilities, signs, and traffic—may increase disorder, thus reducing preference, or they may reflect an increase in the built content, thus depressing preference. The removal or buffering of these elements with more desirable elements, such as foliage, should improve the evaluative image.

Openness

The evaluative maps also point to the importance of openness in the evaluative image (Figure 4.3). In an early study, Lynch and an associate had people walk around a block and record what they noticed (Lynch &

Figure 4.3. Panoramas and Defined Open Spaces Tend to Evoke Favorable Responses
Photograph by Jack L. Nasar.

Rivkin, 1959). Changes in spaciousness or constriction of the streets emerged as a main part of the pedestrians' experience. Subsequent studies have confirmed the prominence of spaciousness and related variables (such as openness, building density, and defined space) in human perception of the environment (Kaplan & Kaplan, 1989; Nasar, 1988b, 1989a, 1994). In his city image work, Lynch (1960) found value in "visual scope" ("vistas and panoramas which increase depth of vision") and defined space ("a strong physical form") as strengthening the effect and memorability of nodes (pp. 76, 106). Research also indicates increases in preference associated with defined openness or open but bounded space. Lynch found "well managed panoramas" as "a staple of urban enjoyment" (p. 44). Research confirms increases in preference associated with openness (Nasar, 1988c; Nasar et al., 1983) and with lower-density development (Lansing et al., 1970). People prefer moderate and defined openness (or some spatial definition) to either wide open or blocked views in both natural scenes (Kaplan & Kaplan, 1989; Nasar, Fisher, & Grannis, 1983; Zube, Pitt, & Anderson, 1974) and built scenes (Hesselgren, 1975; Im, 1984). Observations of behavior in present-day urban areas indicate that many well-known and well-liked plazas have distinctive defined spaces with views to the street. People tend to gather near the edges—where they have a broader and protected view (Whyte, 1980). A study of 192 streets from around the world and dating from the seventh millennium B.C. through 1970 A.D. found enclosed depth and spatial variety as common factors (Rapoport, 1990a).

The preference for open views may arise from perceived status meanings associated with open space, from an increase in perceived order associated with openness, or from the openness itself. We may associate large open spaces with the wealth that enables people to acquire such space. An open view helps the person see and make sense of a scene, whereas a blocked view limits this ability. By making a scene more coherent, the open view may increase preference (Kaplan & Kaplan, 1982). A related theory argues for an evolutionary basis for a preference for open views or prospect (Appleton, 1975). It states that prospect may have enabled our predecessors to see and avoid predators. Those who preferred prospect would have more likely survived than those who did not. In a related matter, the Kaplans put forth mystery (often seen as a deflected vista) as preferred for its promise of new information (Kaplan & Kaplan, 1982, 1989), and others agree that people should prefer such views (Appleton, 1975). Researchers have produced ample empirical evidence that people prefer deflected vistas (Kaplan & Kaplan, 1989). People feeling vulnerable may see a different meaning in a deflected vista (or the promise of further information), however. It may appear to hide a danger such as an attacker. In this case, the hidden information reduces preference. Several studies confirm this hypothesis. Views toward places of concealment evoke fear, and perceived social danger depresses preference for alleys with deflected vistas (Fisher & Nasar, 1992; Herzog & Smith, 1988; Nasar & Fisher, 1993; Nasar, Fisher, & Grannis, 1993; Nasar & Jones, 1997).

The dual meanings of mystery (deflected vistas) conform with broader theoretical views on environmental preference—one seeing preference relating to a balance between making sense and involvement (Kaplan & Kaplan, 1989) and the other seeing preference for an optimal level of uncertainty (Berlyne, 1971). As fear and uncertainty increase (as they do after dark), hidden information becomes incompatible and making sense becomes more important. At the same time, when fear and uncertainty become high, a person prefers environments that reduce the uncertainty. In each case, people dislike deflected vistas.

In sum, the evaluation of openness may depend on context. People may enjoy defined open spaces as well as panoramas of pleasant elements. They may enjoy the first as observers of scenery and the second as potential participants in the space. People also like deflected vistas, except in conditions of uncertainty. For urban design, the research suggests a blending of defined open spaces with panoramas of pleasant elements to create an enjoyable spatial variety. It also suggests the careful design of deflected vistas to give pedestrians open prospect after dark and in other situations of uncertainty.

Historical Significance

As with naturalness, the definition of historical significance rests on the observer's perception of the predominance of historical content (Figure 4.4). Historical content may be authentic or not. If observers consider a place historical, it has historical content to them. Consider the case of architects interviewed about a new building designed in historical style (Marsh, 1993b): They said that they liked it, until they learned it was new. Then they said they disliked it. They liked it when they perceived

Figure 4.4. The Appearance of Historical Significance Tends to Evoke Favorable Responses
Photograph by Jack L. Nasar.

it as historical, and when that perceived meaning changed, so did their evaluation. Nonarchitects did not experience such a reversal.

Certainly, the popularity of restored historical areas in many cities attests to the value we give to history, but several studies support the Knoxville/Chattanooga findings for historical meaning. When researchers asked Parisians what areas they liked best, the Parisians reported "the deepest affection . . . reserved for the central historic areas" (Milgram & Jodelet, 1976, p. 119). Furthermore, when asked where they would take a last walk if they had to go into exile, Parisians most often chose routes in the historic areas. They also often complained about modern apartments and offices replacing the greater charm of the older structures. Other research indicates that these findings are not unique. A survey of Ohio State University alumni about the most-liked building on campus reveals that they most like a historical building for its historical significance (Physical Facilities, 1986). The public also tends to prefer popular or vernacular styles to the "high" styles designed by architects and published in architectural magazines. Stylistic content conveys meanings, with historical content preferred (Verderber & Moore, 1979). One study found respondents to prefer Georgian to modern styles (Whitfield, 1983).

Figure 4.5. Respondents Preferred Farm- and Tudor-Style Homes
Renderings courtesy of Home Planners, Tucson, Arizona, a Division of Hanley-Wood, Inc., 1-800-322-6769.

Across four U.S. cities, using three different sets of housing stimuli and response measures, we found a convergence of preference for vernacular or historical styles over high styles (Nasar, 1989b). Respondents favored farm and Tudor-style houses (Figure 4.5). Interestingly, research in Canada and the United States has found that connotative inferences from dwelling exteriors have consistency across observers and accurately reflect the actual characteristics of the interiors and residents (Duncan, Lindsey, & Buchan, 1985; Sadalla, Verschure, & Burroughs, 1987).

A study of responses to seven buildings found preference related to assessment of age (Marsh, 1993a). Nonarchitects showed consistent preference for a design mimicking a late 19-century style. In describing the reasons for their preference, respondents referred to the mix of visual variety and order. "It keeps your eye busy taking in everything there is to see literally from bottom to top . . . but they're entirely compatible. . . . There's a lot of different stuff, but nothing is at odds with anything else,"

"I like the detail work . . . You can tell the building is not very new—all the detail work" (Marsh, 1993a, pp. 13-14). Architects also initially responded favorably to historical significance, referring to the Greeks, Romans, and Palladio until they realized the building they were viewing was a new building.

Historical significance has another role. Research shows that it increases the imageability of a building as a landmark (Evans, Smith, & Pezdek, 1982; Lynch, 1960), and it may aid recall and wayfinding, especially among the elderly (Evans, Brennan, Skorpanich, & Held, 1984).

People may respond favorably to historical places for one of several reasons. Because historical content enhances building imageabililty (legibility), the preference may relate to improved legibility. Older buildings and areas may have the preferred mix of order and variety. Although each building or element may differ, they tend to fit within a recognizable pattern, going together with a recognizable order. They may also carry meanings—sometimes conveyed with special labels, such as *German Village, Olde Town,* or *Gaslight Square.* Historical significance may also evoke favorable responses through favorable associations, whether accurate or not, with the past or with perceived status. Though the research does not point to a clear explanation, it does suggest that use of historical content enhances the evaluative image.

Order

The Knoxville and Chattanooga respondents said they liked order (Figure 4.6) and disliked disorder. Order has also emerged as a prominent dimension of human response to surroundings. Research consistently finds preferences associated with order and related variables such as organization, coherence, fittingness, congruity, legibility, and clarity (Kaplan & Kaplan, 1989; Nasar, 1989a, 1994; Ulrich, 1983; Wohlwill, 1979). In arguing for legibility, Lynch (1960) recognizes the importance of order. Several studies show organizing variables such as legibility, identifiability, and coherence as important predictors of preference (Kaplan & Kaplan, 1989). In a cross-cultural study of city streets, we found order as a predictor of preference for urban street scenes by both Japanese and U.S. respondents (Nasar, 1984). In studies of housing scenes, we found the ordering variable "clarity" as an important predictor of the evaluative

Figure 4.6. Visual Order Tends to Evoke Favorable Responses
Photograph by Jack L. Nasar.

quality of the scenes (Devlin & Nasar, 1989; Nasar, 1988c). Other work has found preference associated with increases in order in on-site response to buildings (Oostendorp, 1978), increases in order in photographs of architectural exteriors from around the world (Oostendorp & Berlyne, 1978), increases in the congruity of buildings to their natural setting (Wohlwill, 1979, 1982; Wohlwill & Harris, 1980), increases in the compatibility of buildings to neighboring buildings (Groat, 1983, 1984), and increases in the coherence of retail signs (Nasar, 1987). The preference for order may relate to its perceptual character or social meaning associated with orderly environments. Order and upkeep may come together to suggest something about status (Duncan, 1973; Royse, 1969).

Cities can likely enhance their evaluative image through increases in visual order or coherence and through a variety of design features that can help improve perceived order—including legibility, repetition, replication of facade features, uniformity of texture, low contrast between elements or between buildings and their natural context, and identifiability (distinctive elements and focal point; Groat, 1984; Kaplan & Kaplan, 1989; Ulrich, 1983; Wohlwill, 1982). Lynch (1960) talks about the importance of clear structure, differentiation of parts, congruence of form, and use. By changing these features as well as naturalness, openness, upkeep, and historical references, cities may achieve the desired increase in perceived order. Complete order may appear boring and confining, however. People need something to contrast with the order and catch their attention. Complexity can add interest to the order.

Figure 4.7. People Tend to Like Visual Variety
Photograph by Jack L. Nasar.

Complexity

Though the evaluative maps seem to accent order, theory and research indicate that humans also prefer some visual arousal and complexity (Kaplan, 1975; Lozano, 1974; Nasar, 1987; Figure 4.7). We see some indirect reference to this variable in the preferences for historical areas (that tend to have more ornament and detail) and in the visual richness of the most-liked areas. The evaluative maps of Knoxville and Chattanooga, however, show little direct mention of variety or complexity. Like most unplanned cities (Alexander, 1966), these cities (particularly in the CBDs and commercial arteries) may already have a surfeit of visual complexity, and they may have the wrong kind of complexity. Consider a distinction between the visual richness of many different elements and the urban complexity arising from traffic, signs, wires, and poles. The former has some structural order or compatibility among the elements and changes; the latter has less order and higher contrast between the elements. The lack of order found in Knoxville and Chattanooga suggests a need to bring some order or structure to the complexity first. Then the desire for increased complexity (as visual richness rather than chaotic urban elements) might appear.

Complexity and related variables (such as visual richness, ornamentation, information rate, diversity, and variety) have consistently appeared as a prominent aspect of our response to our surroundings (Nasar,

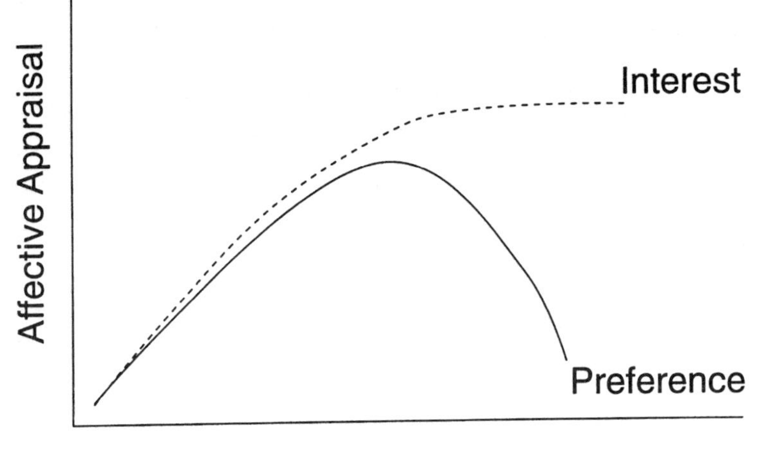

Figure 4.8. Interest and Preference in Relation to Increasing Complexity

1988b, 1989a, 1994). Complexity involves the number of different noticeable elements and the distinctiveness between those elements. Places having few elements or many similar elements appear relatively simple. The introduction of many elements having noticeable differences from one another increases the perceived complexity. Figure 4.8 shows the expected pattern of response to complexity. In theory, interest should increase with complexity. Preference should increase with complexity up to a point, after which preference decreases.

First, consider the findings for interest. Early studies of responses to nonsense figures show exploratory behavior and judged interest to increase with complexity (Berlyne, 1971). Findings for real environments confirm increases in looking time associated with complexity (Wohlwill, 1976). These findings show stability to responses to building exteriors: In on-site judgments of 20 buildings in Toronto, interest related to building complexity (Oostendorp, 1978). A study of response to color slides of architecture found looking time and reported interest related to one another and to building complexity (Oostendorp & Berlyne, 1978). Other work has found increases in interest (or excitement) associated with increased complexity of retail signscapes and housing scenes (Nasar, 1987, 1988c). In sum, the findings agree with the theory that interest and arousal increase with environmental complexity.

The results for preference appear less consistent,[2] but they point to a preference for moderate complexity. Several studies testing nonlinear relationships have confirmed preference associated with moderate complexity. A study of a variety of scenes found preference to have an inverted U-shaped function in relation to independently scaled complexity (Wohlwill, 1974, 1976). Controlled studies also suggest a preference for moderate complexity. In one, respondents evaluated movies of trips through scale-model streets altered to vary diversity: Respondents preferred drives with moderate complexity (Wohlwill, 1976). In another, respondents evaluated a retail signscape varied systematically on complexity and coherence: They preferred the scene having moderate diversity (Nasar, 1987). Both studies confirm preferences for moderate diversity or visual richness. Historical analysis of the forms of streets from around the world through thousands of years also confirms the desirability of complexity (Rapoport, 1990a). The streets share complex enclosing elements, many elements and textures, continual variation in width, and texture and level changes underfoot. In sum, visual richness or complexity evokes interest, and people prefer moderate complexity. Plans that try to provide the desired visual richness in one location and as seen when moving through the city at different speeds enhance the evaluative quality of the city.

Interrelationships, Context, and Contrast

Preference for naturalness, good maintenance, open views, and historical elements may relate to the variables per se, but it may also relate to their effects on perceived coherence (order). Each likable feature may enhance order or coherence, whereas its opposite—intense uses, dilapidation, and restricted movement—may reduce it. Psychologists Stephen and Rachel Kaplan (1989) argue that humans may have a predisposition to favor coherence (or perceived progress toward it) because it has fostered their survival, helping humans make sense of their surroundings so they can act to ensure their safety.

The preferred features may also reflect associations with perceived status. People do notice status and prefer "upper-class" over "lower-class" areas (Duncan, 1973; Lynch, 1976; Michelson, 1976; Royse,

1969). Lynch (1960) notes respondents mentioning avoidance of lower-class areas. People make accurate judgments of social status from environmental cues; these judgments show high levels of agreement among respondents (Cherulnik, 1991; Cherulnik & Wilderman, 1986; Craik & Appleyard, 1980; Duncan, Lindsey, & Buchan, 1985; Sadalla et al., 1987). Americans with more money can better afford larger lots with more space for grass and trees (openness and naturalness), better maintenance (upkeep), design controls for compatibility (order), and either older restored historical homes or new homes that have the detail of historical ones (historical significance). Respondents might judge places with such features as the kind that wealthier persons can afford. Research finds high-status environments associated with some of these features—vegetation, openness, upkeep, and historical significance (Duncan, 1973; Hogan, 1982; Nasar, 1988c; Rapoport, 1990b; Royse, 1969). In a sense, city-scale maps may reflect some stereotypical meanings related to perceived wealth or poverty (social status). From an evolutionary perspective, the preference associated with status may represent a major drive and attraction for the reproductive value of status (Hogan, 1982), or people may simply link status to income and the means to acquire desirable amenities (such as openness, naturalness, and upkeep). Whatever the case, the presence of likable features creates identifiable and sensuous places conveying favorable emotional meaning for many people. If many groups of people of differing ages, stages in life, and social class share the image, it points to directions for improvement.

Where subcultural groups differ, planners can move from the city-scale images to the local image held by residents of neighborhoods. This does not mean that we should overlook other formal qualities that enhance imageability. Planners should attend to the character and arrangement of paths, landmarks, edges, districts, and nodes to enhance the sense of whole and imageability; they should also consider singularity, form simplicity, continuity, dominance, clarity of joint, directional differentiation, visual scope, motion awareness, times series, and names (Lynch, 1960, pp. 105-115).

Urban designers should consider each likable feature in relation to its context. For example, though people like vegetation, this does not make the suburbs intrinsically more likable than cities. The preference may relate to the contrast with less vegetated and more built-up areas. Of course, nature within the urban context is also well liked, and it attracts use (Whyte, 1980), but this may also relate in part to a contrast with built-up areas devoid of vegetation. We should also consider the visual delight that Lynch (1960) finds for urbanized areas in their lights after dark (pp. 18-19). People may notice well-kept areas as preferred in contrast to other areas with signs of decay. The removal of signs and billboards may make other disliked elements, such as poles and wires, more prominent. Large signs or billboards, disliked in Knoxville and Chattanooga, may have a desirable effect in places such as Times Square or Las Vegas. People do not want all places or parts of the city to look alike. They prefer restfulness in some places and excitement and arousal in others (Brower, 1988). People may prefer panoramic views, but the experience may become more vivid when it occurs in contrast with closed areas. Thus, blocked views, though not desirable in themselves, can become an important design element in adding contrast and variety. Historically significant buildings or districts may stand out as preferred in contrast to less varied or less historical places. Rather than treating each feature in isolation, urban designers should consider the context, noticeable differences, and overall organization of features in relation to one another and to contrasting elements to create an evaluative image congruent with its uses.

The general bases for likability in Chattanooga and Knoxville are similar. Respondents tended to select elements that had distinctive forms, visibility, or use significance. Their evaluation of these elements relates to features associated with coherence, perceived social status, and, to a lesser extent, visual variety. The evaluative maps capture many formal and content variables that other research has shown as prominent in human responses to the environment. As people notice and remember these elements, changes in them when handled in relation to the context may produce dramatic changes in the evaluative image.

City Structure and Experience

So far, I have analyzed the evaluative image for areas of agreement. Because the evaluative image develops from an interaction between the environment and observers, differences in either one might produce differences in the evaluative image.

Although Knoxville and Chattanooga have similar bases for likability, the details and structure of the evaluative maps differ. These differences probably result from differences in city structures and experiences. City structure may influence what people notice and remember. Different kinds of observers, such as visitors and residents, use and know the city differently. In each case, the differences affect the evaluative map.

Examination of the maps in each city suggests an influence of city structure. Respondents in both cities like mountains, but Knoxville has views of the mountains and Chattanooga has mountains in the city. Thus, Knoxville observers like places offering views of the mountains, whereas Chattanooga observers like the mountain locations. Other research confirms the importance of relative relief in landscape and shoreline evaluations (Zube, 1980; Zube et al., 1974). In Knoxville, the major paths, which pass by or through the downtown and serve commuters, stand out on the evaluative maps. In Chattanooga, the landmark mountains stand out. The imageability of paths in one city and landmarks in the other probably results from differences in city structure. The results have to do with the environment sampled. If a city has no variance in topography and you test for the predictive power of topography, you will not find any. City structure frames the image.

Comparison of visitor and resident images suggests an influence of different experiences on the evaluative maps. The residents have more extensive, more detailed, and less amorphous images than the visitors. This probably reflects the residents' experience and more detailed knowledge of the city. These findings agree with previous findings for mental maps by Appleyard (1976), Gould and White (1974), and Steinitz (1968). Information also affects the evaluative maps. People tend to know more about places closer to them and places they have experienced (Gould & White, 1974). In Knoxville and Chattanooga, residents cite more places and give clearer definitions of boundaries than visitors. You can see the difference on the evaluative maps of the CBD derived for each group. Residents and visitors also differ in the frequency with which they cite features in need of improvement. Knoxville residents cite industry, highways, utility poles and wires, and the riverfront more often and parking, signs, and billboards less often than visitors.

Though not conclusive, the findings suggest that differences in the maps conform with the view that the evaluative image results from an individual's cumulative experience with the environment. Different kinds of observers and city structures yield differing evaluative images.

Notes

1. Because we did not have people respond directly to scenes in view, the maps show recalled meanings (likes and dislikes), not perceptions.

2. Studies of natural scenes have found preference to increase with complexity (Kaplan et al., 1972; Wohlwill, 1974). The natural scenes may have lacked high enough levels of complexity to obtain the downturn in preference. Studies of urban scenes, which vary more widely in complexity, have also found preference to increase with complexity (Devlin & Nasar, 1989; Herzog et al., 1976; Kaplan et al., 1972; Nasar, 1988c). These findings appear flawed in other ways. Some studies left content (such as land use) uncontrolled or examined preference across content categories. Because the relationship between complexity and preference varies across content categories (Herzog et al., 1976, 1982; Kaplan & Kaplan, 1989; Wohlwill, 1974), the findings may result from the content associated with complexity rather than the complexity itself. Some studies had the same respondents rate complexity and preference so that the rating rather than complexity may have affected preference. Independent judgments of complexity reversed the direction of the relationship between preference and complexity and no significant linear relationship emerged (Kaplan & Kaplan, 1989). Finally, some researchers have neglected to test for a nonlinear relationship.

5

Evaluating the Method

In examining the concept of urban likability, we interviewed a sample of people about their likes and dislikes in the city. Through telephone interviews with city residents, we obtained oral descriptions of liked and disliked elements. Through in-person interviews with city visitors, we obtained oral and written responses in relation to maps of the city. Several questions arise about the value of the methods: (1) How useful or practical are they? (2) How accurately do the interview questions gauge what they intend to measure? (3) How consistent are responses to the measures? and (4) How well do the results generalize to the real-world situations to which they are intended to apply? The first question focuses on usefulness. The others deal with *construct validity, reliability,* and *external validity.* The studies in Knoxville and Chattanooga sought to gain a preliminary view of the public image of the cities and to derive some directions for design for improving community appearance. They centered on exploring urban likability and developing hypotheses about the factors associated with it. We did not design the studies to address questions about the validity and reliability of the methods explicitly: We can look at other research to evaluate reliability and validity. This chapter first describes the method in detail and then describes related findings on reliability and validity. It also discusses possible improvements on the methods for the future.

The Method

We sought a diverse sample of both residents and visitors in each city, interviewing 160 residents and 120 visitors in Knoxville and 60 residents and 60 visitors in Chattanooga. We interviewed residents and visitors because of likely differences between their images due to differences in familiarity and experience and because of the potential value to the local economy of understanding visitor (tourist) reactions.

For both residents and visitors, we sought to achieve a representative sample. Interviewers contacted residents by phone. To select resident respondents, we listed all three-digit residential exchanges in each city. To these, we added four numbers at random until we had the desired sample size. Interviewers dialed each number three times (between early morning and late evening). If no one answered, interviewers replaced the number with a new random selection. They interviewed between five and ten adults at each exchange. More than 90% of those contacted agreed to participate.

Interviewers contacted visitors in person in hotels and motels. In Knoxville, we selected hotels and motels at random in each of six zones of the city: east, west, north, south, downtown, and the university. We conducted no more than four interviews at any one site. In Chattanooga, we selected six hotels and motels in six different parts of the city and interviewed 10 persons at each site. In both cities, interviewers approached the first passerby they saw in the public space of the lodge. If the interviewer was rejected, or on completion of an interview, the interviewer approached the next passerby. Tables 5.1 and 5.2 show the sociodemographic characteristics of the samples and related census data for each city (U.S. Department of Commerce, 1983).

The interview included five groups of questions:

1. We asked respondents to identify up to five areas they considered the most pleasant visually and to identify five areas they considered the most unpleasant visually. We varied the order so that about half the interviewers asked about likes first and the other half asked about dislikes first. Residents responded orally over the phone. Visitors responded in person with the aid of maps.
2. We asked for the boundaries to each area. To delineate the boundaries, we probed for street names, landmarks, known buildings, and other elements. For the in-person interviews, we used maps to assist in this process.

TABLE 5.1 Distribution of Respondents by Sociodemographic Characteristics: Knoxville

	1980 Census	*Residents Interviewed (n = 160)*	*Visitors Interviewed (n = 120)*
Gender			
Male	46%	39%	56%
Female	54	61	44
Race			
Black	10	11	9
White	89	88	87
Other	1	1	4
Age			
Under 20	27	19	3
21-39	35	49	54
40-60	19	20	30
Over 60	19	11	7
No response	—	1	6
Income			
$0-$9,999	30	42	9
$10,000-24,999	46	38	58
Over $25,000	24	9	28
No response	—	13	5

TABLE 5.2 Distribution of Respondents by Sociodemographic Characteristics: Chattanooga

	1980 Census	*Residents Interviewed (n = 60)*	*Visitors Interviewed (n = 60)*
Gender			
Male	47%	48%	67%
Female	53	52	33
Race			
Black	19	17	20
White	80	83	80
Other	1	—	—
Age			
Under 20	30	8	12
21-39	31	58	58
40-60	21	22	14
Over 60	19	12	14
Income			
$0-9,999	28	15	8
$10,000-24,999	45	17	17
Over $25,000	27	21	42
No response	—	47	33

3. We asked respondents to name the physical features that accounted for the evaluations.
4. In Knoxville, we asked people to select from a list of elements the one most in need of visual improvement. The list included industry, highways, signs and billboards, buildings, utility poles and wires, riverfront, parking lots, buses, railways, service stations, and other. The in-person interview form had space for respondents to list other features. Phone interviewers had residents describe other features if they reported "other." The use of a fixed choice of elements allows a formal comparison among the elements, but it may yield misleading results. The list may not capture categories relevant to respondents (even with the "other" category). It also mixes levels of environment (such as industry and signs). In Chattanooga, we used an open-ended question: "If you could change one thing, what would you change to improve the appearance of Chattanooga?" The open-ended question overcomes the limitation of the fixed-choice procedure, but it can hinder the objective analysis. It forces the researcher to rely on the discretion of someone other than the respondent to group responses into categories for tallying.
5. Finally, the interviewer asked for the sex, race, age, and income of each respondent.

We informed respondents of the study purpose. To reduce biases in response, the interviewers told respondents that there were no right or wrong answers and that we wanted only their honest opinions.

From each interview, we developed one evaluative map for each respondent. We overlaid groups of five maps on one another and tabulated and mapped the frequency of overlap in areas. We combined these summary maps again in groups of five until we had the composite.

Usefulness

This particular approach may have more value in identifying problems than in developing innovative solutions because it depends on respondents' present knowledge of what exists in the city rather than their speculations about what could exist. Other methods that have people respond to unfamiliar scenes or think of other cities can help uncover innovative solutions less tied to people's local experience. The development of the composite maps by overlaying maps on one another proved somewhat imprecise. For greater precision, the Geographic Information Systems could be used to help code responses from each individual's map into a grid of cells across the city, tally the scores for each cell to derive preference scores, tally reasons given for responses in each cell, link the results to other social and physical conditions in the cell, and transfer the results into map form.

The method did prove useful. The relatively short interview enabled us to gather the necessary information from a broad sample of respondents fairly quickly. Longer interviews could include other questions to refine the description of the evaluative image or the respondents. Each interview took about 15 minutes, and respondents appeared quite animated and interested in the interview. They also answered quite rapidly, suggesting that they had the knowledge about meanings readily available.

Validity and Reliability

In each city, we sought samples with geographic and social diversity. Recall that the amount and type of information an individual has about places in a city vary with that individual's location (where he or she lives, travels, and works). Gould and White (1974) point out some of these locational influences on knowledge and preference: Individuals tend to exaggerate the local area about which they know more, and they tend to have lower emotional involvement or preference as the perceived distance rises; the information a group of persons has about a place increases with the size of the place's population and decreases with the square root of the subjective estimate of the distance to the place. Thus, studies in two cultures, Sweden and Nigeria, found similar equations showing that information related to population divided by the square root of distance. Psychological and physical barriers can affect perceived distance. Just as a river or train track may physically divide a city, so might social factors such as ethnicity, race, and religion psychologically divide a city. Consider cities such as Quebec, Belfast, Beirut, and cities in the former Yugoslavia. Many U.S. cities have pronounced racial and income boundaries. Variations in social factors create barriers to information. Studies of regional maps confirmed sharp declines in preference for some areas because of an information barrier. In England, the view from Bristol across the channel to the north shows a sharp decline in preference because travel to these areas is a long and circuitous trip. In Sweden, children recall places in their own country but few nearby places in Norway, whereas in Norway the reverse pattern occurs. The boundary affects the flow of information for children quite close together in physical distance. Research in Tanzania shows strong influences on preference related to cultural groups, such as Arabs and Masai, living in particular areas. In Malaysia, different preference maps emerge for Malaysian and Chinese residents relating to the cultural attitudes of each group (Gould & White, 1974). Because the location of observers, their distance from places, the size of those places, and physical and social barriers in the information space can affect preference maps, a sample that overrepresents one group or part of a city might skew the evaluative map toward preferences for local areas. For both the resident and visitor surveys, we sought geographic and social diversity.

None of the samples had enough respondents to allow precise statistical estimates of the confidence limits and percentage of the population. To have a confidence level of 95% that a particular percentage of the population liking a particular area is accurate plus or minus 5%, cities the size of Knoxville or Chattanooga would require a sample of 384 residents. To increase the confidence level to 99%, the sample would have to increase to 663 residents. If we assume that each city has a few hundred thousand visitors each year, similar sample sizes would apply to visitors.

By randomly selecting phone numbers and using repeat calls, we achieved a fairly representative sampling of residents in each city. A researcher can reach more than 92% of households by phone (Frankel & Frankel, 1987). Though phone surveys may underrepresent some groups such as the poor, transient, and socially isolated, all surveys tend to underrepresent these groups (Groves, 1987). Compared with census samples, the Knoxville and Chattanooga samples had a lower percentage of respondents under 20 and over 60 years old and a higher percentage of respondents between 21 and 39 years old. Knoxville also had a lower percentage of males than appears in the census. The income distribution of the samples remains uncertain because of nonresponses to the question about income. I have less confidence in the representativeness of the visitor sample, which consists of visitors available in public areas in hotels or motels and willing to participate. It may underrepresent visitors not staying in such lodgings or not frequenting public areas. The groups may not have different evaluative images of the city, however.

Due to potential limits of sample size and representativeness, the derived maps may not depict the "true" evaluative image of each city. For that, we might need a retest with larger and improved samples. Still, we interviewed large enough and diverse enough samples to uncover consistencies in response and relative levels of likes and dislikes for places. The correlation of the findings with other research findings suggests that they may generalize to the true public image.

The interviews relied on people's recall for areas of the city they had experienced. Do such responses reflect on-site experience? Research suggests that they might. In one set of studies, one group of observers rated the perceived safety for eight areas shown on a map, and another group rated the perceived safety for the same areas during actual visits (Nasar & Fisher, 1993). Comparisons of these ratings show strong correspondence between the map ratings (from memory) and ratings during the actual visit. Both sets of ratings also have strong correspondence with unobtrusive observations of pedestrian behavior: Pedestrians avoid the feared areas. Furthermore, studies elsewhere of physical features related to maps of recalled fear and to reported fears in a walk-through show similar features to those obtained for the eight areas (Nasar, Fisher, & Grannis, 1993; Nasar & Jones, 1997).

The fear studies looked at relatively small-scale familiar places. Do the results apply to larger-scale or less familiar places? Data at the city scale suggest that recall does apply to on-site response. Lynch (1960) found that verbal interviews about memorable elements in the city show strong similarity to responses to photographs and to responses of passersby at the actual sites. The verbal recall generalizes well to on-site response. He also found one difference between responses through verbal interviews and sketch maps: sensitivity. Verbal interviews retrieve more elements than sketch maps. Although some respondents had a low correlation in their responses on the two methods, the composites from the sketch maps and the verbal interviews agreed. At least for recall, it appears that verbal interviews and, to a lesser extent, sketch maps reflect on-site experience with the city.

Other research also suggests that the evaluative image at the city scale may generalize to on-site experience. In Vancouver, researchers compared ratings of a map with ratings obtained on-site for 16 areas (Snodgrass & Russell, 1986). They found that the map evaluations did generalize to evaluations during actual visits. The two sets of ratings had similar scores reflected in a high correlation between the mean ratings.[1]

Additional tests can clarify how well responses to various modes of experience apply to the real day-to-day experience of the city. This might involve small follow-up studies of on-site response to examine the accuracy of responses obtained through other means. For a general knowledge base, this might also require several studies comparing responses from various simulations with responses on-site. It should also recognize that distortions in people's knowledge about a city may affect their evaluative map (Milgram et al., 1972). If most people in the sample have not experienced a particular area, the area will not appear on the map. Though the appearance of this area may not matter in the evaluative image of the full city, it may matter to local observers. It helps to think of the evaluative image as a nested hierarchy. The city image combines the shared images of various neighborhoods and districts that, in turn, combine the shared images of blocks in the areas that, in turn, combine the private images of each building owner or resident. As the scale and imageability of the elements decrease, local images and meanings take on more importance than broader public images. To capture the relevant information, planners might go beyond the city-scale maps to obtain evaluative maps of neighborhoods or districts as seen by residents of those areas and evaluative maps of blocks as seen by block residents. Studies might also consider different cities and levels of familiarity to identify areas of accuracy and inaccuracy and to identify ways to obtain more accurate results.

For applying the results, planners may also want to know the extent to which findings for one group apply to others. Research suggests that they do. Recall the evidence of strong consensus in preferences across various groups. Other research supports these findings. In Vancouver, researchers obtained responses from two groups of college students, one at the University of British Columbia and the other at Langara College (Snodgrass & Russell, 1986). Though the students lived in different parts of the city, the ratings of the areas by the two groups showed high correlations.[2] Because students may share similar images relating to their shared age, interests, and social orbit, the findings may not tell whether other subgroups share similar evaluative images. Another city-scale study explicitly compared responses of a broader sample of respondents. Steinitz's (1968) study of the intensity, type, and significance of places and activities found no substantial differences between groups. He found no differences between inner-city or outer-city residents, no differences between automobile commuters and mass transit commuters, and no differences between observers of different socioeconomic class. In agreement with our findings for visitors and residents, Steinitz did find changes related to length of residency. Newcomers, who are presumably more similar to visitors than to residents, have less complex knowledge of the city. Kang (1990) compared responses to different house styles across five social class groups. He found no differences in preference across all groups, except for one: The designers differed from the others.

In spite of general agreement on the image, locational and experiential differences of respondents of different social classes, ages, stages in life cycle, gender, or primary mode of travel may lead to some differences as well (Lynch, 1960). The evidence so far points more to differences in spatial knowledge rather than meaning across various groups. A study of Los Angeles found that upper-class whites had a wider and more detailed mental map of the city than lower-class blacks, who in turn had a less restricted map than a Spanish-speaking minority (Orleans, 1973). Other research confirms ethnic differences in mental maps (Appleyard, 1976; Francescato & Mebane, 1983; Magana, 1978; Maurer & Baxter, 1972). In one study that at first appears to run against expectations, Appleyard (1976) found that lower-class residents in Ciudad Guayana have greater spatial knowledge than the upper class. A closer look shows that the differences relate to experience. The lower class have longer trips to work than the upper class, who work near their homes.

Stage in life cycle, from teenager to young single to couple to parents with young children to parents with older children to elderly empty nesters, involves changes in the use of the environment and social orbit (Michelson, 1976, 1987), and these changes may affect the content and evaluation of the evaluative image. Research indicates that children go through developmental stages in their mental mapping abilities and that the elderly may have some deficits in their maps (Evans, 1980). Males and females use different areas and notice different things, but most research on cognitive maps has found few if any sex differences, except for gender-based restrictions on home ranges (Evans, 1980). Girls with more restricted home ranges than boys produce smaller and less-accurate sketch maps (Hart, 1979). Research also shows differences in spatial knowledge related to the mode of travel (Appleyard, 1976; Beck & Wood, 1976).

In each case, differences in a group's experience with the city may affect its information field, thus influencing the evaluative map. Factors such as the observer's location, social class, stage in life cycle, sex, or mode of travel may affect the person's experience of the city. This influences the information people have. If it influences their interpretations of it, it may also affect the content of the evaluative map. Knowledge of some potential group differences could have value for planning cities in which different groups inhabit different areas or experience different parts of the city from one another. In selecting possible candidates for such differences, planners should first look for likely group differences in movement through space.

Now let us look at the consistency (or reliability) of responses. If a person's responses to questions about likability or the evaluative image change when the individual and the environment have remained stable, that raises questions about the reliability of the measurement scale.

The research in Vancouver examined the consistency of the ratings across the original 23 raters (Snodgrass & Russell, 1986). Looking at the degree to which each rater gave similar ratings to each other rater, the researchers found high consistency in the ratings.[3] The testing revealed the sample of 23 raters as adequate to achieve consistency for the ratings of pleasure. Studies at a regional scale found that interobserver agreement on likes and dislikes increases with age (Gould & White, 1974). Adult respondents exhibit a high level of agreement on preferences. This pattern shows stability in studies of two other cultures—Swedish and Nigerian.

The findings suggest the likely accuracy and reliability of evaluative maps for adult respondents. We need further study examining for a diverse sample the responses of the same persons on separate occasions to confirm the reliability (or reproducibility) of the verbal or sketch map methods. A literature review on mental maps confirms that sketch maps have consistently shown adequate reliability, however, and that free recall (the method used here) exceeds the recall shown on sketch maps (Evans, 1980).

Refining the Method

The method of deriving evaluative images, though efficient, may have some shortcomings. I have already discussed the need to consider larger and more representative samples and reliability. Beyond that, several decisions depend on researcher discretion. In helping respondents establish boundaries for the areas, in defining areas when respondents did not give clear boundaries, and in grouping reasons for evaluations into categories, we had to make discretionary judgments. One could obtain a more objective aggregate of the data using nonmetric multidimensional scaling techniques or computer mapping (Baird, 1979; Golledge, 1977), though some subjectivity may arise during the interview in defining boundaries. The researcher faces a trade-off between open-ended responses and providing the respondent with cues and landmarks. The first approach might obtain unclear or ambiguous boundary definitions. The second might lead the respondent. The studies of Vancouver, Tokyo, and the university neighborhood suggest another way to deal with boundaries. Establish well-known districts and their boundaries in advance. Then obtain responses to the districts. Be mindful that this approach captures responses to districts only. It may miss evaluative responses to other elements, such as paths, edges, nodes, and landmarks.

Maps derived from open-ended questions about areas liked and disliked appear reliable. The studies in Knoxville and Chattanooga used careful follow-up questions about the elements (whether they were streets, rivers, railroads, buildings, edges, nodes, or districts) that define the boundaries. One could supplement this verbal information with maps and photographs of places in the city to help respondents delineate areas and boundaries. One could partition the city into small cells to allow the coding of likability scores at a finer grain and more precise mapping by computer. This approach would also allow the coding and mapping of individuals' intensity of preference. In addition, an independent detailed field reconnaissance, such as those described by Lynch (1960) or Steinitz (1968), could provide information on the physical character of the liked and disliked areas. One could supplement this with census data describing the sociocultural characteristics of the residents. The field inventory and census information could also help delineate areas and boundaries, and the use of computer coding could reduce researcher discretion in defining boundaries.

Further exploration of methods for defining reasons for likability is needed. We used open-ended questions for their efficiency and to avoid leading respondents in any predefined direction. This approach gets at overt aspects of the evaluative image. By emphasizing physical features, the open-ended questions may have missed some other factors such as assumptions about status, sociocultural factors, and meanings. They also may have missed some important latent features affecting people's evaluation. People may not be aware of or may not express certain elements that affect their evaluative response. The open-ended responses depend on researcher discretion in grouping reasons. Future research could supplement or replace the open-ended responses with a comprehensive checklist of attributes. Such a list might encompass all sense modalities and attributes that include physical, sociocultural, natural, built, fixed, semifixed, and nonfixed elements (Rapoport, 1993). Observers would use the list to check off, rank order, or score the bases for their evaluations. Although it would lengthen the interview, a checklist derived from earlier open-ended responses and research findings could broaden the range of potential reasons and make quantification easier and more objective. A systematic evaluation and comparison of the two methods could clarify the relative merits of each approach.

The applicability of various methods to various scales of environment should be evaluated. Some methods work better for one scale of environment, whereas others work better for other scales. In a pilot study of open-ended evaluations of a three-block commercial strip, we examined responses to photos, responses to maps, written (verbal) responses, and responses on-site. We found similarities and differences related to method. Responses to maps tended to emphasize buildings and streets labeled on the map. This captured a recalled image biased by labeling. The open-ended verbal (written) responses with no visual cues tended to refer to general descriptions, such as trees, trash, or signs, capturing a general memory of the whole area. By emphasizing recall, the map and verbal responses may have missed some aspects of direct experience of the area. Responses to photographs tended to refer to a mix of specific features, such as a specific bench, trash can, bike chained to trees, street vendor, plaza, or building, and general features, such as trees or signs. Observers did not have to rely on their recall or mental image of the area. They referred to what they saw in the photos. Responses on-site varied depending on where the observer stood along the strip. The responses also referred to specific features, such as sidewalks, flowers, trash, and graffiti, and general features, such as trees and signs. As with photos, the on-site response captured the direct experience but may have missed some aspects of the recalled mental image. When combined, the three off-site methods did capture many of the elements reported on-site. Both the recalled image and the direct image have importance: The one may affect behavioral plans whereas the other may affect direct experience in a place. A multimethod approach, each method with unique biases, would probably converge on the broader evaluative image. Still, we need further study for different scales of environment evaluating the merits of the various modes of response (on-site evaluations, evaluations obtained using photos, hand-drawn maps, open-ended verbal responses, responses to verbally labeled districts, and responses to districts labeled on maps).

The Knoxville and Chattanooga studies simply asked people what they liked and disliked. The studies in Vancouver, Paris, and Tokyo suggest that interviewers could ask observers to report several dimensions of their emotional appraisals and the intensity of those appraisals. Other work suggests the importance of meanings such as status. Though these studies obtained such responses to predefined districts, planners could obtain similar responses in an open-ended format. One could pick an emotional appraisal (such as exciting or dull) or a meaning (high status or low status) and ask respondents to delineate areas as high or low on that particular scale. Then, with a complete map, one could ask respondents to rate the intensity of feeling for each area on the selected scale. Observers could do this for their own maps or at a later point for a composite map. The procedure could be repeated for other relevant aspects of emotional response and meaning.

Though research has identified a good list of descriptors for emotional meanings—pleasant-unpleasant, arousing-sleepy, and calming-distressing—we need similar kinds of research to offer guidance for the relevant dimensions of social meanings. Status, identity, and friendliness may represent three aspects of sociocultural meaning worth consideration, but others aspects may also apply.

To determine areas of agreement on city appearance, we aggregated responses across the city. This technique may produce maps that reinforce average values at the expense of important values of small groups. Given the wide and varied populations of many cities, planners may want to consider two other approaches. One might obtain evaluative images from various sociocultural, "taste" groups. By analyzing commonalities and contrasts in these images, planners could identify ways to improve the evaluative image without sacrificing unique preferences of various groups. For smaller-scale areas, such as a neighborhood, commercial district, or pedestrian street, planners might study the evaluative image of the local area by local residents and passersby. The visual form could then be programmed for the people most affected, the local subculture.

So far, I have discussed *static* images obtained by people evaluating an imagined or directly experienced scene. None of the methods refers explicitly to the *dynamics* of moving through a city. Static and dynamic images differ. Moving through a city, whether by car, bus, train, or bike or on foot, the traveler experiences a sequence of views. The experience changes with speed, important barriers, and points of decision (Appleyard, Lynch, & Myer, 1964; Rapoport, 1977; Rapoport & Hawkes, 1970; Thiel, 1995). Different cues and different levels of complexity become noticeable with changes in speed and distance. Thus, pedestrians and motorists have different experiences—with the motorist noticing large-scale, coarse-grain elements and the pedestrian noticing smaller-scale, finer-grain elements. Similarly, at a distance from a building, place, or skyline, people tend to notice coarser-grain features such as shape, height, or average color. As people move closer, smaller-scale and finer-grain elements such as surface texture or ornament become more noticeable.

Because certain roads may represent gateways, cities may want to evaluate and improve the image these roads convey to visitors. City structure may affect the imageability of gateways, but gateways may include roads from the airport to downtown hotels and convention centers, transit connections to major destinations, roads from highway exits into the downtown or business centers, or connections to other frequently visited destinations, such as stadiums, theaters, and museums. In some cases, the experience of walking through an airport to a connecting flight may leave visitors with an evaluative image of a city.

In each case, observers experience the evaluative image sequentially as they move through the environment. This kind of experience might require different methods. For the dynamic evaluative image of the view along a road, a researcher might take people on field trips along the route to discuss their likes and dislikes along the way. Because getting a representative sample of visitors to take such a trip might be impractical, the researcher can bring the environment to visitors. A convenient approach would involve bringing the environment to observers through their imagination. Ask visitors or residents to imagine taking a particular trip. Lynch (1960) did something like this to identify recalled elements along a route. He asked people for "complete and explicit directions for the trip that you normally take going from home to work" (p. 141). Then he asked them to imagine taking the trip and to describe the experience.

Slight variations on Lynch's (1960) script can help a researcher get at the evaluative image. Research could focus on highly used common routes, such as the route from the airport. It could use both a field reconnaissance and public interviews. The field reconnaissance would record the prominent physical features of the trip. Researchers have developed notation systems for describing the physical features of a trip as seen at ground level (Appleyard et al., 1964; Thiel, 1995). Then, with a variation on Lynch, the public interviews could ask persons to

> describe *and evaluate* the sequence of things you see, hear or smell along the way . . . that have become important to you. We are interested *in your emotional reactions to and the meanings you see in the physical surroundings.* It is not important if you can't remember the names of streets or places. (Lynch, 1960, p. 141, italics mine)

The script might continue as follows: At what points do you feel excitement, pleasure, calmness, arousal, boredom, displeasure, distress, or sleepiness? Where do you notice changes in status? What is it about the things that make you feel that way? We are interested in the physical features that may have led you to respond in the way you did. (The interviewer would probe to determine specific locations for the various reactions.)

For a more direct experience of the trip, one can use color film, videotapes, or slides of the trip. The simulations would have to present a wide enough view to allow the observer the choice of looking ahead or glancing at details alongside the route (Appleyard et al., 1964). With the loss of continuous movement, multiple slides at various points along the trip represent a relatively easy option. At a higher cost, one could use several motion picture cameras to create a wide-angle motion view. Hypermedia possibilities could enable the observer to look at various scenes along the way. Planners can obtain responses from people in the role of passenger or the role of driver. Instructions to "drivers" might tell them to imagine driving along the road focusing on the road and other vehicles as they would when driving. The researcher could then tape-record the observers' running commentary about what they notice and what kind of emotions they feel. By timing the audiotape to the visual presentation, one can determine what on the trip evokes the viewer's response and link it to the field reconnaissance. With a replay, an interviewer could request the reasons for the various emotions expressed or intensity of the emotion. A few people might be taken on actual trips to confirm the accuracy of the responses to simulation.

To form a composite out of the record of individual trips, the analysis could tally the frequency with which respondents mention various elements, the frequency with which they associate each of those elements with various emotional appraisals, and the frequency with which they cite various physical features as influencing their appraisals. The researcher could link this to the physical reconnaissance of the trip. The results, in numeric or graphic (evaluative map) form, would depict the shared evaluative image of the trip. They could serve as a basis for a visual plan to improve the gateway evaluative image. As the area changes, new appraisals can provide updates and evaluations of the changes.

Cities do not have dynamic images through space only. They also have dynamic images through time. The cityscape changes with the seasons, from day to night, and over the life course of citizens. The evaluative image may develop and change, and it may vary in relation to external conditions of season and lighting. The studies of Knoxville and Chattanooga overlooked potential variations in the evaluative image over time, but this aspect has relevance for both theory and practice of urban design. For a visual plan, one needs to know what aspects of the maps remain stable across seasons or lighting conditions and what aspects change. This suggests the need for small follow-up studies of responses under different conditions, with specific reference to those conditions. Findings may vary with the physical conditions of the particular place, but to some extent there may be commonalities in ways in which responses in one condition apply or do not apply to other conditions.

The evaluative image may also change and develop over the life course of residents. Planners may gain knowledge of changes in the image related to experience from the study of the evaluative image of persons who have lived in the city for various amounts of time and from the study of the evaluative image of persons at various points in their life course—young children, teens, young adults, couples, couples with children, empty nesters. One could get this information from an evaluative image study by obtaining responses from the relevant groups and the relevant sociodemographic information to allow the classification. This would give a time slice comparison of the images of the various groups. A richer alternative would track developmental changes in evaluative image over time with the same persons.

Finally, one must consider scale of the environment. I referred earlier to the evaluative image as having a hierarchical form. Just as when you plan a trip to a distant destination you may first think about the broad aspects of the trip (such as the highway from one city to another), then about smaller-scale details (the roads that move from the highway to a particular area), and then about still smaller scale details (the particular road that leads to a particular house), so the evaluative image has a hierarchy of information to which people may refer. People have images of their region, their city, their neighborhood, their street, their house, and portions of their house. The detail of the evaluative image will likely increase as the size of the study area decreases. Maps of neighborhoods may uncover more specific elements than maps at the city scale. At different times and for different purposes, different evaluative images come to prominence. Planners should use this knowledge to tailor the evaluative image to the particular problem at hand. The broad visual plans for a city may vary for specific districts or neighborhoods. Visual plans for a neighborhood may draw largely on neighborhood residents' evaluative images. Plans for a major artery or commercial strip on the edge of the neighborhood might draw on the evaluative image of passers-by. The evaluative image for the central business district might draw on a still broader sample.

I see the use of evaluative mapping as a dynamic process. Communities can assess their evaluative image to derive directions for the future—defining what they should preserve, what they should add, what they should change, and what they should remove (Jones, 1990). Then, as they put the plan into action, the evaluative image can chart the effects of a planning intervention. Did the intervention produce the expected results? For example, New York City enacted regulations requiring new development in the Times Square area to have large brightly colored signs and billboards. Had the city obtained an evaluative image prior to the policy, it could have implemented the policy and then tested the evaluative image to assess the effectiveness of the policy. The follow-up evaluative image study can also apply to a specific imageable feature (such as a large new building at a highly visible location) to see its effect on the image. To get at likely effect prior to development, planners could obtain evaluations of the proposed development in simulation form. Through the continuing assessment of the existing and desired evaluative image as the city and its population changes, communities can guide plans and designs to improve the quality of the image and the quality of life for the present and the future.

Notes

1. For pleasure, the ratings from recall and on-site had a correlation of .93 ($p < .001$). For arousal, the two sets of ratings had a correlation of .97 ($p < .001$).

2. For pleasure, the scores of the two groups of students correlated at .91 ($p < .001$). For arousal, they had a correlation of .83 ($p < .001$).

3. The Cronbach interrater reliability score for pleasure indicated high reliability (.95). For arousal, it had a lower score, still indicating fair reliability (.74).

6

Other Dimensions of the Evaluative Image

The studies of Knoxville and Chattanooga examined one aspect of the evaluative image—adult preference. Different populations, scales of environment, or aspects of meaning might reveal different evaluative images. This chapter examines other aspects of the evaluative image. It includes studies of the following:

1. Children's evaluative images of a small town
2. The evaluative image at different scales of environment, including a neighborhood and commercial strip
3. Other dimensions of emotional meanings
4. The image at various times of day and night

Sixth Graders' View of a Small Town

With a population of approximately 29,000 residents, Gilbert, Arizona, is a small town made up predominantly of single-family housing, parks, and a small town center. As a preamble to updating the general plan, Scott Anderson, director of planning in Gilbert, interviewed 330 sixth-grade children in nine local elementary schools, obtaining approximately 100 evaluative maps (Anderson, 1991). The planners gave the students a one-hour training session on the history of planning and the background of a planner. Then they put them into groups of five or six and asked them to indicate on a map

1. What things did they like in Gilbert?
2. What things did they dislike in Gilbert?

Note that the interview did not specify visual appearance. It asked for a more general description of likes and dislikes. From the individual maps, planners created a composite map and a list of liked and disliked elements in the community.

The planners classified places into three levels of imageability: *most imageable* (cited by more than 90% of the students); *imageable* (cited by between 60% and 89% of the students); and *fairly imageable* (cited by between 30% and 59% of students). As in Knoxville and Chattanooga, imageability did not necessarily translate into likability.

The seven places classified as *most imageable* included two places neither liked nor disliked (a McDonald's and an Eat-a-Burger) and one disliked place (downtown). The 10 places classified as *imageable* included two places with neutral evaluations (library, post office) and six disliked places (four schools, a railroad, and the dairies). The nine places classified as *fairly imageable* had two places with neutral evaluations and seven disliked places. The meaning and evaluative quality of the imageable features define the evaluative image of the city.

The findings do not necessarily mean that imageability and likability are independent. The evaluative descriptions of Gilbert also show a link between the two. The more imageable places tend to have higher likability than the less imageable places. Table 6.1 shows the mean likability scores and proportion of likable places for each level of imageability.

As seen in Table 6.1, the more imageable places have higher mean likability scores and higher proportion of likable places than places lower in imageability. A statistical test of differences between the mean likability scores reveals a statistically significant difference between the three levels of likability and a sizable effect.[1] The small number of places in each category suggests caution in interpreting this test. These data suggest that either Gilbert conveys a favorable image to the students or the students tend to recall or report more liked places than disliked ones.

TABLE 6.1 Likability of Imageable Places in Gilbert, Arizona

	Least Square Mean Likability Score (1= least liked to 5 = most liked)	*Proportion Liked (mean > 3.5)*
Place category		
Most imageable (n = 7)		
(Cited by 90% of children)	3.864	57%
Imageable (n = 10)		
(cited by between 60% and 89% of children)	2.764	10%
Fairly Imageable (n = 9)		
(cited by between 30% and 59% of children)	2.366	0%

Nine schools appeared in the student ratings. Student scores showed three as particularly disliked for physical cues that may have associations with social meanings. Students reported disliking these schools because they looked too old, needed repair, and had vandalism—all potential cues to undesirable social conditions. In contrast, the students reported liking some newer schools because they looked new and had more space. Some students complained about their plain look, however.

What accounts for the imageability and likability in Gilbert? Recall the role of use significance, distinctiveness, and visibility in imageability. Gilbert does not have direct measures of these features, but an inspection of the imageable feature suggests that for the children in Gilbert, use significance and visibility dominate. The *most imageable* places include restaurants, recreation areas, and downtown. These places all have heavy use and visibility in the community. To a lesser extent, some may have distinctiveness: "The look is different." The second level of *imageable* places also appears to have use significance and visibility. It includes elementary schools, the library, the post office, and stores. Physical form and meaning play a role in some of the most cited or often cited places. For the downtown, students also mentioned the lack of form that would cause them to frequent it. Freestone Park (a 50-acre regional park; see Figure 6.1) drew comments about its water, open space, and landscaping that made its form unique. The railroad stood out for its meaning, as blocking progress because of its location. Students referred to the canal system more for the color of the water than for its trail links. They cited

Figure 6.1. Residents Liked Freestone Park, Rated 4.7 out of 5 (fun, good play area, peaceful, nice landscape, a place to get away, safe, pleasant, clean air, relaxing, open spaces)
Photo courtesy of W. Scott Anderson.

Figure 6.2. Residents Liked the 7-11, Rated 4.34 (good hangout, convenient, looks good)
Photo courtesy of W. Scott Anderson.

the old town hall for its appearance. One place stood out for its distinctiveness in a nonvisual sense: Students cited the dairies for their bad smells.

Although the Gilbert study did not ask explicitly about visual quality, students often referred to community appearance. They gave general comments about appearance. On the favorable side, they mentioned such things as "looks good," "nice area," "like the way it looks," "good design," "pleasant," "relaxing," "inside looks nice," and "cool building." On the unfavorable side, they mentioned such things as "dislike," "it's boring," "ugly building," "ugly," "looks bad," "buildings are boring," "tracks are ugly," "does not look good," and "looks like a prison." They also referred to specific elements that they liked or disliked. For Knoxville and Chattanooga, I suggest five kinds of preferred features—naturalness, upkeep/civilities, openness, order, and historical significance. I also refer to complexity (visual richness). When the Gilbert students mentioned specific physical elements underlying their likes or dislikes, they often referred to features that fit into five of the six categories: naturalness, upkeep/ civilities, openness, history, and visual richness. Though they did not often explicitly mention order, many features that they liked or disliked may be related to perceived order and coherence.

1. Naturalness: Students referred favorably to water, nice landscaping, a nice park. For example, they liked the canal system for the "color of the water," and they gave negative comments on the lack of greenery—no landscaping, no grass, or needing grass.

2. Upkeep/civilities: Students gave favorable mention to good upkeep, mentioning such things as "cleanliness," "clean air," "clean," and "nice landscaping" (also a reference to naturalness). They gave negative evaluations to perceived incivilities, referring to "smells," "pollutes the air," "needs repair," "too old," "too much vandalism," "water too dirty," "water smells," "turns green," "homes not maintained." For example, they disliked the town hall as "trashed," "too old," and "needs repair." In one explicit reference to symbolic meaning, they described it as "no longer representing Gilbert." They disliked one school because they said it had "too much vandalism." The references to "clean air," "smells," "water smells," and "pollutes the air" point to nonvisual features, suggesting the potential value of considering a variety of attributes across different sense modalities—visual and nonvisual (Rapoport, 1993).

3. Openness: Students gave favorable mention to openness, referring to "big," "wide open space," "not overcrowded." They gave negative evaluations to crowding and constriction, referring to places as "crowded" or "too small."

4. Historical significance: Students gave some favorable comments on history, saying things such as "like culture" or "represents town history."

5. Visual richness (complexity): Students gave favorable comments to uniqueness and complexity, saying things such as "unique to Arizona" or "unique" and "look too plain," "needs more colors," "buildings are boring." They said they disliked the "plain" look of some of the newer schools. Colors also played a role in their preferences, though the specifics remain unclear. They mentioned good colors, nice colors on buildings, ugly colors.

Because the interview asked about general likes and dislikes rather than appearances, the planners found some additional categories having to do with use and activities. The imageable elements tended toward activity nodes, but the Gilbert results showed preferences for each of Lynch's (1960) imageable elements. Students mentioned likely

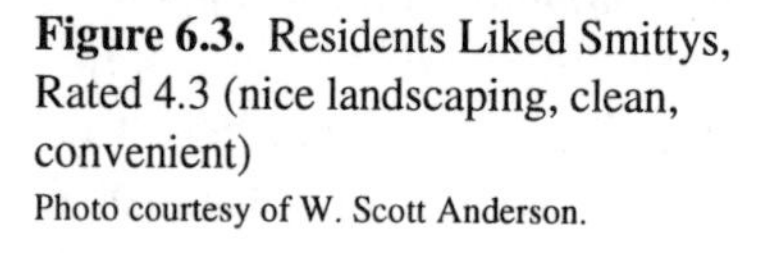

Figure 6.3. Residents Liked Smittys, Rated 4.3 (nice landscaping, clean, convenient)
Photo courtesy of W. Scott Anderson.

Figure 6.4. Residents Liked Ski Spring, Rated 4.1 (good entertainment, fun sport, unique in Arizona, water, good recreation)
Photo courtesy of W. Scott Anderson.

nodes (such as McDonalds), landmarks (such as the post office), paths (Guadalupe Road), edges (such as the canals and the railroad), and districts (the downtown). The prevalence of nodes may result from the questions asked, the structure of the town, or the students' emphasis on activity.

You may have noticed the finer grain of information for Gilbert compared with that of Knoxville and Chattanooga. In this small town, students referred to specific and smaller buildings (McDonald's, Eat-a-Burger, the library), whereas in Knoxville and Chattanooga, the adult respondents referred to larger elements. As the size of the rated area becomes smaller, information takes on a finer grain. Thus, evaluative maps of smaller towns tend to have details not present in evaluative maps of larger cities. For the large city, the shared image comes at a coarser image, more broad and general. Similarly, assessments of a district or neighborhood would show even finer-grained information.

An Inner-City Neighborhood

German Village covers 233 acres of land directly south and contiguous to the central business district of Columbus. It has a population of between 4,000 and 5,000 residents. The city of Columbus designated German Village as a historical district in the 1960s. Residents have

Figure 6.5. Typical German Village House
Photograph by Jack L. Nasar.

renewed more than 1,600 buildings since then. In 1975, the area was placed on the National Register of Historic Places. It has a grid pattern of streets, a major park, and a major church that should stand out as a landmark and node. It has relatively small lots (30′ x 90′) and house sizes (17′ x 31′). Houses are relatively close to one another and the street. Typical houses have a rosy-brick wall, limestone foundation, limestone lintels, and sandstone sills above and below the windows (ornamentation), a gable roof, and a wrought iron fence (see Figure 6.5). The area has a mixture of uses, with retail uses at various points in the district and concentrated along a section of Third Street, the major road through German Village. This inner-city neighborhood has many features favored for neotraditional development.

Selecting households at random, we interviewed a diverse sample of 60 residents of German Village. We asked them what one thing they would change to improve the visual quality of the area and what one thing they would keep for its visual quality. As in Knoxville and Chattanooga, the stress on visual quality may have eliminated certain nonvisual features from responses. We also asked residents to identify what areas they liked (and why) and what areas they disliked (and why). Figure 6.6 shows their evaluative map of the district.

As expected, the neighborhood-scale map captures finer-grained entities. For example, when asked what they would change to improve

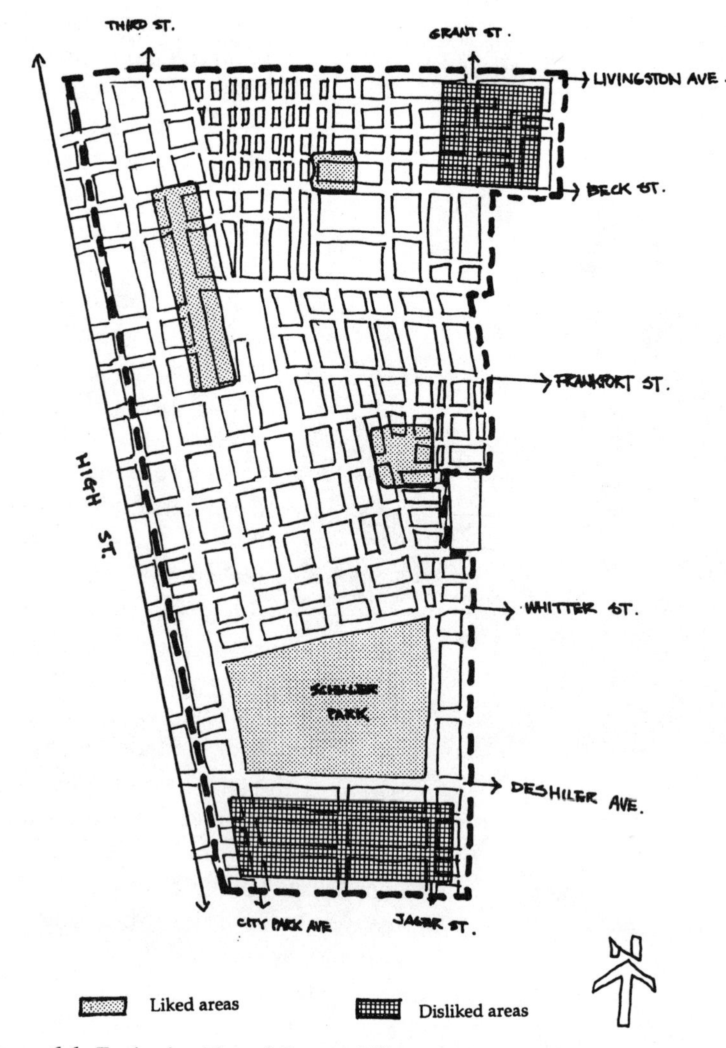

Figure 6.6. Evaluative Map of German Village (Columbus, Ohio)
Drawing by Jack L. Nasar.

visual quality, residents often referred to the sidewalk conditions and landscaping. They also mentioned reducing traffic. They said they liked a small park, a larger park, some small shops, and a church. They said they disliked two areas, each occupying between four and six square blocks. These finer-grained entities have meaning to the neighborhood residents.

Figure 6.7. Residents Like St. Mary Church
Photograph by Jack L. Nasar.

Their evaluative image stresses features that observers might see as landmarks, nodes, and districts more than paths and edges. It includes a church (landmark), two parks, several shops (nodes), and two small areas (districts) on the north and south edge (Figures 6.7, 6.8, 6.9, and 6.10).

As we have found elsewhere, this reflects a preference for naturalness, upkeep, openness, order, historical significance, and visual richness (complexity). Respondents said they like the parks for their naturalness and openness and as community gathering places. They said they dislike two areas primarily for their rundown appearance and like other areas for their cleanliness (upkeep, perceived status). Order, historical significance, and visual richness come out in comments about what they most want to keep.

Figure 6.8. Residents Like Schiller Park
Photograph by Jack L. Nasar.

Figure 6.9. Residents Like This Small Commercial Area
Photograph by Jack L. Nasar.

Figure 6.10. Residents Dislike the City Park Neighborhood
Photograph by Jack L. Nasar.

They most frequently refer to the appearance of buildings. The buildings have historical significance, and through design controls they have both order and visual richness.

Multiple Meanings in Vancouver

Beyond preference, evaluative meanings also include excitement, relaxation, social status, identity, and friendliness (Nasar, 1989b; Rapoport, 1990b; Ward & Russell, 1981). Preference may encompass the other aspects of emotional meaning. Thus, Gould and White (1974) report that evaluative maps of the United States derived from open-ended responses that the investigators classified into categories and tallied produce virtually the same results as those obtained from preference ranking. The open-ended responses categories collapsed into one scale—preference. Though other meanings relate to preference, each may also relate to different places and characteristics of place. Maps of these meanings may differ from one another and from maps focusing on pure preference (likes and dislikes).

In Vancouver, psychologists Jacalyn Snodgrass and James Russell (1986) explored other aspects of emotional meaning. They derived a list of perceived districts in the city and then obtained evaluations of each district. They had 23 students at the University of British Columbia (UBC) identify areas they perceived as districts in the city. Though student responses may not typify the responses of adult residents, the study showed one way to map several dimensions of emotional meaning. The researchers first asked the students to draw borders around and label each area on a map. This produced a list of 52 districts. Later, the researchers displayed the map as a large slide. Pointing to each district on the slide, they had the students rate how well each of eight descriptors described each area.[2] The descriptors included *pleasant, exciting, arousing, distressing, unpleasant, gloomy, sleepy,* and *relaxing*. Examination of the scores for the various districts revealed that the scales differentiated among the 52 areas. The researchers reported statistically significant differences in the scores across the districts.

Using a statistical procedure to group the eight scales, the researchers found that the scales grouped into two main factors: pleasure and arousal.[3] From the scores on these two factors, the researchers classified each district into one of eight evaluative categories: pleasant, unpleasant, relaxing, distressing, arousing, sleepy, exciting, and gloomy. Districts with neutral scores on arousal were classified as either pleasant or unpleasant depending on their pleasantness scores. The researchers also classified the intensity of the emotion on each scale. So, for example, the districts could vary in their degree of relaxingness or degree of excitingness. Figure 6.11 shows a map of the 52 Vancouver districts for a variety of emotional appraisals.

Different evaluative maps are associated with each emotional axis. These maps give planners additional information to use in shaping the future appearance of the city. In contrast to Knoxville and Chattanooga, the predominance of the pleasant shading in Vancouver shows the students evaluating the city, for the most part, as pleasant. Areas judged as exciting, pleasant, or relaxing may represent areas worth preserving, whereas areas judged as stressful and unpleasant point to potential areas for change and improvement.

The findings show that cities can have different kinds of favorable or unfavorable meanings. People might want excitement in some places and times and relaxation in other contexts. It may seem difficult to translate poetic metaphors such as relaxing, distressing, exciting, or gloomy into explicit directions for design, but research can systematically study the relationships between such evaluations and the physical and sociocultural aspects of places to identify the kinds of features to achieve the desired emotional quality. Studies have already done so for *relaxing, exciting,* and *distress* (Nasar, 1987; Nasar & Fisher, 1993; Nasar & Kang, 1989). The Vancouver studies did not look for the attributes associated

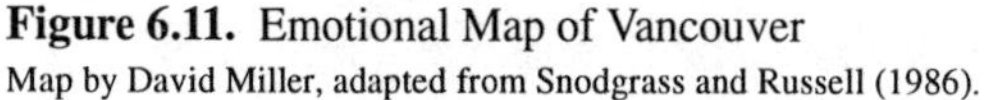

Figure 6.11. Emotional Map of Vancouver
Map by David Miller, adapted from Snodgrass and Russell (1986).

with the meanings, nor did it look at social meanings such as perceived status. To do this, a study could have respondents rate the social status or the friendliness of each area, and it could ask them about the features associated with such responses. As an alternative, researchers could obtain responses to photographs, systematically altering one attribute at a time to see how the changes affect various meanings, as did an early study looking at social inferences from physical cues (Royse, 1969).

Multiple Meanings in Tokyo

In Tokyo, psychologist and planner Kazunori Hanyu (1993) had students respond to 23 well-known wards. The 32 students, familiar with the wards, rated the similarity between each possible pair of wards. For people

TABLE 6.2 Tokyo Adjective Pairs Grouped by Factors

Factor (Percent of Variance)		
Scale		
Static–dynamic (50.39%)	**Evaluation** (26.44%)	**Other** (11.91%)
Coarse–smooth	Good–bad	Heavy–light
Relaxed–tense	Beautiful–ugly	Exciting–unexciting
Warm–cool	Rich–poor	
Sharp–blunt	Wide–narrow	
Fast–slow	Huge–tiny	
Strong–weak	Deep–shallow	
Active–passive		

SOURCE: Hanyu (1993).

Figure 6.12. Evaluative Map of Tokyo
Courtesy of Kazunori Hanyu.

TABLE 6.3 Student Evaluative Image of Tokyo

Evaluation	*Location (wards)*	*Most Frequently Reported Places and Meanings*
Bad	Shitamachi	Various rivers (historically muddy, polluted, and flooding)
(Unpleasant and dull)	(Adachi, Arakawa, Edogawa, Katsushika, Kita, Kohtoh, Sumida, Taitoh)	A sightseeing town (declining but many shrines and temples)
		An older zoo with caged animals
		A declining area to the north
		A discount market
		Higher density development
		Traditional Japanese-style landscapes
		Old temples and shrines
		Lower town
		Landfill
		Coarse
		Drab
		Far from center
		Famous residents or characters
Good (Pleasant and relaxing)	Yamanote (Ohta, Meguro, Nakano, Setagaya, Suginami)	Denen-chofu (A garden city); *denen-chofu* came to mean excellent & expensive land
		Tyuoh line (subway that runs to suburbs with pleasant images)
		Residential
		High-class residential
		Nihon University (interview site)
		Mid-rise complex (has rock concerts)
Static (Sleepy)	Between Shitamachi and Yamanote on north (Itabashi, Nerima, Toshima)	Older, small-scale amusement park
		An apartment complex
		Farms and radish produced in Nerima
		An entertainment area
		Oldest high-rise building and one of the tallest in Tokyo (has aquarium, shopping, and offices)
		A hospital
		Far from center

continued

familiar with an area, earlier research had found that responses to the name of a place yield responses similar to those obtained from photographs of the place (Herzog, Kaplan, & Kaplan, 1976). From the similarity data, Hanyu derived five clusters of wards. He then had 87 students evaluate each ward on 15 7-point adjective scales (see Table 6.2). Note in Table 6.2 that the student responses might not generalize to responses

TABLE 6.3 *Continued*

Evaluation	*Location (wards)*	*Most Frequently Reported Places and Meanings*
Dynamic (Arousing)	Between Shitamachi and Yamanote on south (Minato, Shinagawa, Tiyoda, Tyuoh)	Imperial Palace A classy/expensive hotel Roppongi (a newer entertainment area with top-ranked restaurants, shops, boutiques, and theaters) Ginza town (expensive sushi and tempura restaurants, department stores, and Kabuki theater) Akasaka (like Ginza but also known for Geisha girls) Tukiji (a famous and busy fish market) Fashionable, rich Port, the sea Business district Administrative office area Central
Moderately dynamic	Bunkyo (north center)	Academic town Domed stadium (Tokyo Giants) Tokyo University
Exciting	Center (Shibuya, Shinjyuku)	Kabuki-cho (large entertainment area with many small restaurants, bars, and sexual entertainment) Hachiko (famous statue/meeting place) High-rise buildings Entertainment quarters Fashionable For young people Crowded Shopping

SOURCE: Hanyu (1993).

by other groups. As did Snodgrass and Russell (1986), Hanyu (1993) looked at the groupings of the 15 scales to identify groupings of similar scales. He found that the scales grouped into two factors—evaluation and static–dynamic.[4] These factors echo the pleasure and arousal factors of Vancouver, with evaluation resembling pleasure and static–dynamic resembling arousing.

Relating the evaluative ratings to each ward cluster showed differences in the evaluative images of the clusters. Figure 6.12 shows the resulting evaluative map of the wards in Tokyo. In it, one can see that the observers judged one ward cluster as unpleasant, another as pleasant, one as static, another as arousing, another as moderately arousing, and one as exciting. None of the wards received a distressing rating, possibly because of the lower crime and fear of crime rate in Japan.

Table 6.3 summarizes the places and meanings associated with the evaluative responses. To understand the evaluative groupings, one should understand a common bipolar concept of social meaning that Tokyo residents use to distinguish areas in Tokyo: Yamanote-Shitamachi (K. Hanyu, personal correspondence, July 23, 1996). From the 14th century through the 19th century, the Samurai lived in the Yamanote (or mountainside) area and the artisans, crafts workers, and merchants lived in the Shitamachi (or lower town), also a business area. The areas have different population groups and patterns of land use. Yamanote has upper-middle-class residents and newer, less dense residential development. Shitamachi has lower-middle-class residents and older, denser, mixed-use development, including commercial development.

The land use, age of the area, and social class distinction between Yamanote and Shitamachi may account for much of the pure evaluative dimension. The evaluation axis varied from wards in the Shitamachi area to wards in the Yamanote area. The unpleasant and dull cluster included eight traditional lower-middle-class wards corresponding to the Shitamachi area, which has come to mean middle- to low-class, crowded, mixed-use, and more traditional Japanese residential areas. It has old temples and shrines, traditional Japanese-style landscape, high-density land use, and some commercial areas (Figure 6.13). Students reported associating these northern areas with poor rice farmers (from farther north) who use a station in one of the areas to enter Tokyo.

In contrast, the pleasant and relaxing cluster included five upper-middle-class wards corresponding to the Yamanote (mountainside) area, which has come to mean high class, quiet, and residential. These wards also include some middle-sized commercial areas on the western side of Tokyo (Figure 6.14). The residential areas have relatively lower densities and larger lot sizes. The unpleasant and pleasant clusters are located on opposite sides of Tokyo, with unpleasant in the northeast and pleasant in the southwest. The two areas also differ in land value and crime rate, with higher land value and lower crime in the pleasant Yamanote area.

Wards on the static arousing axes varied from remote residential areas to centrally located business and governmental areas. This dimension relates to patterns of land use—residential (static) versus office-commercial

Figure 6.13. Scene from Unpleasant Group
Courtesy of Kazunori Hanyu.

(arousing) —that vary in activity level. The static cluster (located between the Shitamachi and Yamanote areas) has three middle-class residential wards that include a mix of residential, commercial, and industrial uses (Figure 6.15). The land value and crime rates for this cluster fall between those for Shitamachi and Yamanote.

Opposite this, the arousing cluster has four wards consisting of the original central business district. Overlooking Tokyo Bay, this office-commercial section of the Yamanote area houses major governmental and public facilities—such as the Imperial Palace, the House of Parliament, and the Supreme Court—as well as high-rise office buildings (Figure 6.16). This

Figure 6.14. Scene from Pleasant Group
Courtesy of Kazunori Hanyu.

Figure 6.15. Scene from Static Group
Courtesy of Kazunori Hanyu.

Figure 6.16. Scene from Arousing Group
Courtesy of Kazunori Hanyu.

cluster has the highest land value of all the clusters.

The students classified one ward—Bunkyo—as moderately arousing. It has Tokyo University (the highest-ranked university in Japan), many publishers, and the first domed stadium in Japan and home of the Tokyo Giants, the most popular baseball team in Japan. Tokyo residents recognize the area as an academic area: Its name means "academic town." The static and arousing areas are located on opposite sides of Tokyo, with the static to the northwest, the arousing to the southeast, and the moderately arousing in between.

The exciting cluster (an office-commercial part of the Yamanote

Figure 6.17. Scene from Exciting Group
Courtesy of Kazunori Hanyu.

area) has two newer city centers with a mix of high-rise offices, retail, and entertainment (Figure 6.17). It has a lower land value than the older arousing government-office cluster. It is located both spatially and conceptually between the arousing and the pleasant districts.

Newcastle: Day and Night

People experience their cities at various times during the day and night. As Lynch (1976) points out, activities and perceptions of parts of a city vary over the course of a day. Some areas have continuous use, others empty at night, others become especially active at night, and others shift from day to night activity. The evaluative maps so far overlook changes in experience through the course of the day and night, or they rest on the assumption that the map merges the observer's various modes of experience. Geographers Don Parkes and Nigel Thrift (1980) report a study that looked at time-specific images of a city. They assigned 120 students (ages 19 to 22) into six groups of 20. One group responded to time-free images, similar to those we have discussed earlier, whereas each other group received a specific time to consider: mid-morning (10:30 a.m.), mid-afternoon (3:30 p.m.) of a working day, early evening (8:30 p.m.), and late night (2:00 a.m.).

A pilot study had identified a list of 25 well-known points in the city center and surrounding area. Parkes and Thrift (1980) had each respondent describe and evaluate each place on ten 9-point bipolar scales: feel at home–feel strange in, like–dislike, beautiful–ugly, visit often–visit seldom, new–old, work–recreation, cultural–commercial, quiet–noisy, busy–idle, safe–unsafe. The respondents could add personal constructs and rate them on 9-point scales as well.

The results indicated a similar image structure and meanings for the two daytime periods, but unique image structures and meanings for early evening, late night, and the time-free images. As the time shifted from daytime to early evening, the image shifted from active (familiar) to inactive (unfamiliar) and from disliked commercial state to a better-liked cultural-recreational state. Thus, the students responded more favorably to the more active cultural-recreational conditions of early evening than to the less active commercial conditions during the day. At late night, another change occurred along a third dimension: the image of most elements shifted toward lower levels of activity, familiarity, and safety. A few images reversed from safe and pleasing during the day to unsafe, dangerous, and frightening in the late night.

Parkes and Thrift (1980) describe the changing appearance of the city as darkness approaches: "lights switch on and shine from buildings . . ."; "the major roads appear as rivers of light" stopping intermittently at traffic lights; "some spaces . . . have a higher proportion of their buildings with lights switched on than others"; "then as clock and social time passes lightened areas darken and darkened areas brighten" (p. 131). Although Parkes and Thrift do not examine the elements associated with the changing meanings over time, other studies do.

A Neighborhood: Day and Night

Hanyu (1995) explored the daytime and nighttime evaluative images of a neighborhood in Columbus, Ohio. The neighborhood, the University District, covers 2 square miles with a population of approximately 45,000 households. It has a daytime population of 110,000, the second-largest

TABLE 6.4 Five-Point Scales Used for the University Area, Columbus, Ohio

Evaluative scales (stressing observers' emotional appraisal of the scene)
- Pleasant–unpleasant
- Exciting–boring
- Relaxing–distressing
- Fearful–safe
- Interesting–uninteresting
- Active–inactive

Environmental descriptor scales (1 = not at all, 5 = a great deal)
- Complexity
- Legibility
- Coherence
- Mystery
- Openness
- Typicality
- Naturalness
- Brightness
- Uniform lighting
- Nuisance elements
- Vehicles
- Familiarity

SOURCE: Hanyu (1995).

daytime population in Columbus, exceeded only by the central business district.

Hanyu (1995) divided the University District into 20 subareas such that streetside scenes in each area had similar physical characteristics to one another but differed from such scenes in other areas. The divisions grew from earlier work by a neighborhood planning group, census data, and walks through the neighborhood. The subareas differed from one another in socioeconomic, housing, and physical conditions. The average subarea covered 8 square blocks, with most covering fewer than 5 square blocks. Three larger areas covered 31, 24, and 21 blocks. Within each area, Hanyu took a daytime and nighttime photograph of one representative residential scene. He shot the daytime and nighttime photos from the same places and with the same viewing angles.

Hanyu (1995) had 52 students evaluate the scenes shown in the photographs. Students voluntarily participated in the study as part of a capstone course. Fewer than 5% of the potential participants declined

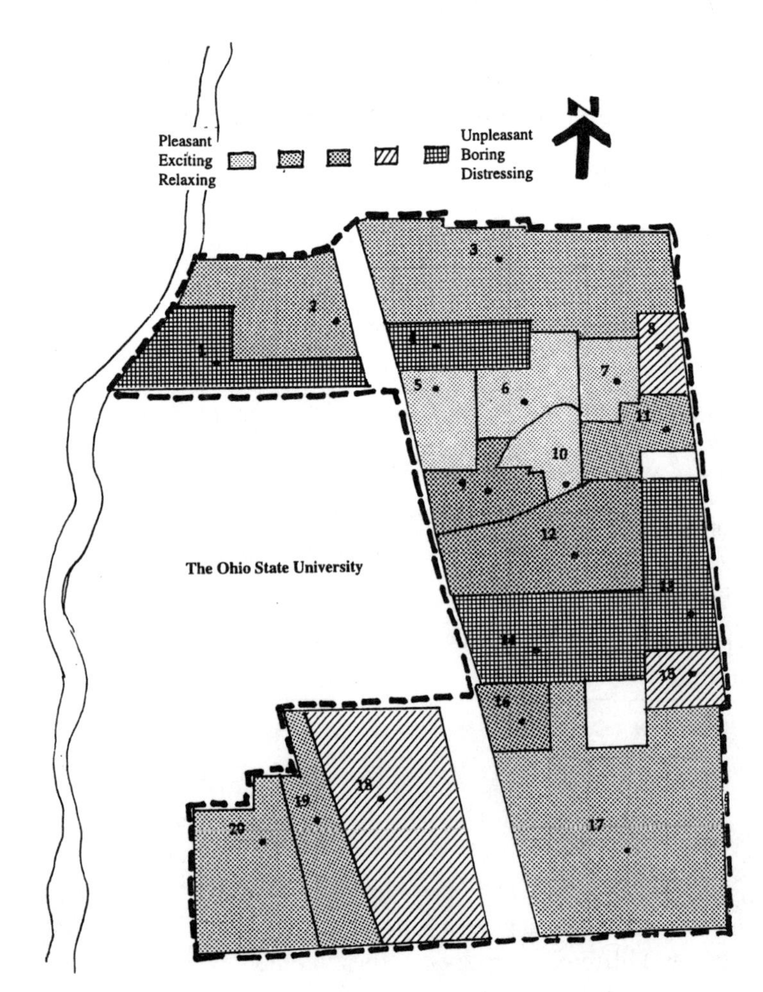

Figure 6.18. Daytime Evaluative Map of the University Area, Columbus, Ohio
Drawing by Jack L. Nasar.

participation in the study. The sample included more than 20 different majors. Participants had an average age of approximately 25 years old (with a range from 20 to 46 years old) and an average tenure in Columbus of 6 years (with a range of from half a year to 24 years). Students were assigned at random to one of the two conditions—daytime scenes or nighttime scenes. Twenty-four students (15 males and 9 females) evaluated the daytime scenes, and 28 students (20 males and 8 females) evaluated the nighttime

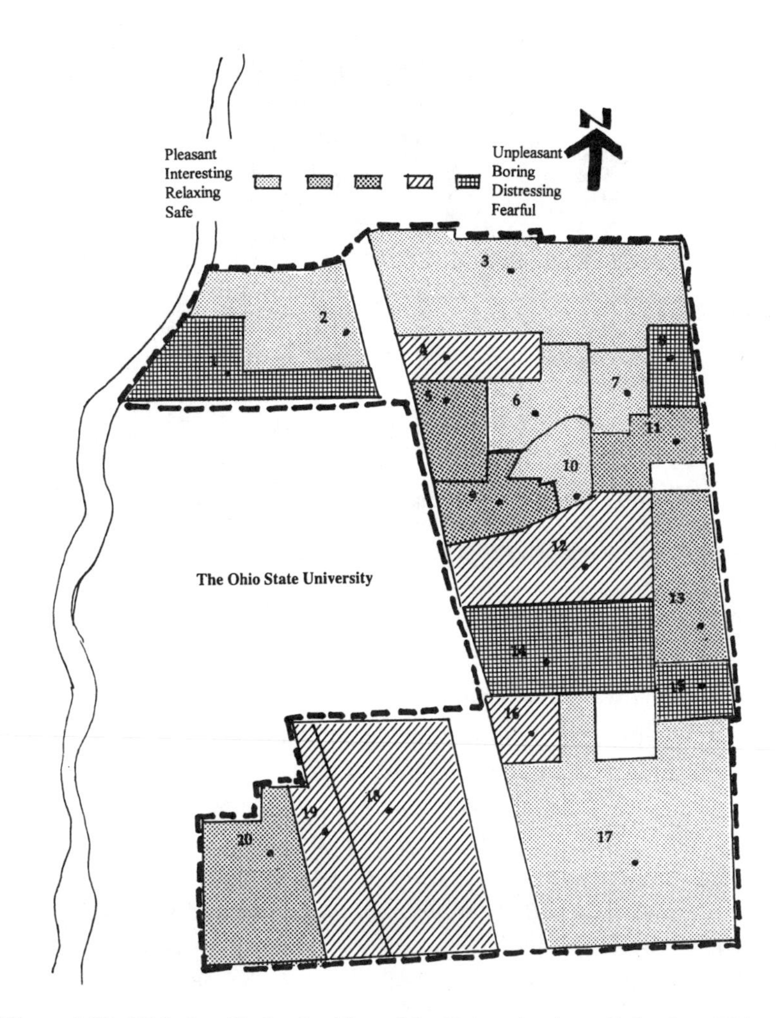

Figure 6.19. Nighttime Evaluative Map of the University Area, Columbus, Ohio
Drawing by Jack L. Nasar.

scenes. They sorted the scenes (as 3" × 5" color photographs) for similarity, and they rated each scene (shown as a slide) on 18 5-point scales. Twelve scales described physical characteristics of the scenes and six described evaluative meanings of the scenes (Table 6.4).

For the daytime data, the study (Hanyu, 1995) found a prominent evaluative dimension that combines pleasant, exciting, and relaxing. For nighttime, four scales—pleasant, relaxing, interesting, and safety—combine to make a similar emotional dimension. The analysis reveals agree-

Figure 6.20. Scenes from Liked Areas in University District
Photographs by Jack L. Nasar.

ment in the scores across the scales. Figures 6.18 and 6.19 show the resulting daytime and nighttime evaluative maps, and Figures 6.20 and 6.21 show some of the liked and disliked areas.

As you can see, the evaluative ratings differentiated between the arcas for both the daytime and nighttime evaluations. The day and night evaluative images have surprising similarities, however. Though respondents gave lower scores to the nighttime scenes,[5] the day and night scores have a high correlation and maintain a similar order.[6] Most areas received similar evaluations in both conditions. The maps placed areas into one

Figure 6.21. Scenes from Disliked Areas in University District
Photographs by Jack L. Nasar.

of five categories, from pleasant to unpleasant. Nine of the 20 areas had scores in the same category on each map, and eight more had scores in adjacent categories on each map. The scores for all these areas showed similar directions in day and night. Two other areas showed reversals, however: Area 19 had favorable scores during the day and unfavorable scores at night, and area 13 had unfavorable scores during the day and favorable scores at night.

Hanyu (1995) also looked at the relationships between independently obtained ratings of physical characteristics and the emotional quality of the scenes. The evaluative images for both day and night showed favorable evaluations associated with increases in naturalness, openness, and complexity.[7]

For the daytime data (Hanyu, 1995), the results show that naturalness and openness predict overall preference.[8] Disorder predicts arousingness.[9] Legibility predicts behavioral activity.[10] Visibility predicts perceived safety.[11] Built nuisance elements predict distress.[12] Similar findings emerge for nighttime: The nighttime results also show overall naturalness and openness as predicting preference,[13] disorder predicting arousal,[14] and visibility predicting perceived safety.[15]

The study (Hanyu, 1995) also suggests a status component of meaning. Both the day and night conditions distinguish single-family units from multifamily units, rating the latter as less pleasant and more fearful. The pattern of findings for each kind of housing type separately echoes the findings for the full set: Increases in naturalness, openness, and visibility enhance the perceived pleasantness and safety of the area. Increases in order and legibility and decreases in disordering elements, such as vehicles, poles, wires, signs, trash cans, deterioration, mystery, and complexity, reduce the perceived arousal of the area.

As did Parkes and Thrift (1980), Hanyu (1995) found important differences in the day and night maps. Though relations between the evaluative and descriptor scales remained relatively stable, the ratings of particular areas varied from day to night. For example, areas 9 and 19 received favorable ratings during the day and neutral to unfavorable ratings after dark. Areas 13, 14, 18, and 20 received less favorable ratings during the day than after dark. The findings suggest that the factors influencing meaning may vary with context (Rapoport, 1993).

This, in turn, points to the need for additional comparative research to broaden the base of evidence. The differences do not necessarily imply different processes underlying the responses. A similar process may explain the apparent differences. Additional comparative study of various kinds of experience (such as time of day, season, land use, and subcultures) can help refine concepts to understand cause better. For policy, the potential differences related to context (such as night and day) suggest the need for context-sensitive design guidelines and solutions that work for the populations affected in the various contexts. For example, the different daytime and nighttime evaluations of a deflected vista (hidden information ahead) may result from a difference in uncertainty and compatibility of the form with the person. Guidelines sensitive to both conditions might call for a deflected vista during daylight but the use of permeable vegetation and lighting to remove concealment after dark.

Students represented appropriate observers for this study (Hanyu, 1995) because the neighborhood is primarily a student one. Numerous other studies showing little differences between groups suggest that the results might apply to other residents, but without interviewing other residents, one cannot know how well the student assessments apply to others. Though the neighborhood-scale map takes on a finer texture than the coarse city-scale map, two of the city-scale factors in preference emerge at the neighborhood scale: naturalness and openness. Perhaps the others (upkeep, order, and history) do not vary extensively in the neighborhood studied.

A Commercial Strip

Now we drop down to the scale of a street. We looked at the evaluative image of a three-block portion of a commercial strip. One expects to find finer-grain impressions than those at the neighborhood or city scale. We interviewed 60 people, 45 off-site and 15 on-site. To get respondents, we stopped every third person encountered, alternating sex. More than 95% of those approached agreed to participate. The sample varied in sex, background, and age from 19 years old to older than 41, though most described themselves as students (Table 6.5).

Dividing the sample into four groups of 15, we used four different interview techniques: verbal recall, responses to maps, responses to color photos, and on-site evaluations. Although the different methods showed agreement, they also picked up some differences across measures, suggesting a need for a closer look at measures and the need for the use of multiple measures.

TABLE 6.5 Characteristics of Respondents to Commercial Strip

Characteristic	*Percent (n = 60)*
Gender	
Male	55%
Female	45
Age	
18 years old	3
19-25	52
26-41	33
41+	12
Background	
Student	67
Work in area	15
Live in area	8

Visual nuisances stood out as the most mentioned disliked elements. Disliked by most respondents (80%), these elements include trash, signs, bikes, wires, traffic, news racks, and poles. Of these elements, respondents referred most often to trash and signs. Respondents also gave some site-specific responses, referring negatively to a specific street vendor, an older bookstore, and a clock on the side of a building.

Naturalness and openness dominated the liked elements, though upkeep and history also emerged as liked. Most respondents (60%) referred to trees and plazas, with slightly more references to trees than plazas. A smaller group cited two well-kept buildings, one older and historical in character and the other more modern. As with the dislikes, some preferences referred to small elements such as benches or trash cans. The composite evaluative images suggest a need for improved upkeep, increase in vegetation and plazas, and decrease in visual clutter (such as signs).

In another study (Nasar, 1987), we looked specifically at the evaluative image of retail signs in the same area. The results show the desirability of moderate complexity and high coherence in the signscape. They also show excitement related to increased complexity and decreased coherence. We manipulated the amount of complexity and coherence of signs in a scale model of a commercial strip and photographed nine simulated signscapes showing three levels of sign complexity by three levels of sign coherence. Interviews with potential shoppers revealed increased excitement associated with increases in complexity and decreases in coherence. Preference and desirability increased with coherence and peaked for moderate complexity with high coherence. Respondents reported that they would most like to visit, shop, and spend time in an area such as the one with moderate complexity and high coherence. Separate interviews with merchants revealed that they most wanted their commercial area to resemble the one with high coherence and moderate complexity.

Summing It Up

I have discussed evaluative images for a small town, two relatively small cities, and two relatively large cities. I have discussed evaluative images for cities from different cultures (U.S., Canada, Australia, Japan). I have examined evaluative images of children, college students, adult residents, and visitors. I have looked at the evaluative image from the scale of a city to the scale of a neighborhood and a small commercial strip. I have looked at the image for various times of the day. I have looked at recall and direct perception of evaluative images. I have moved from one-dimensional preference maps to maps capturing several dimensions of emotional experience.

The studies show some differences. The specific kinds of element cited (landmark, path, edge, district, or node) vary with the context and are related to the variability and presence of these elements in the area evaluated. For the element to stand out as imageable, the area has to have those elements, and it has to have noticeable differences in them. Some meanings vary from daytime to night. Different areas, sociocultural features, and physical features relate to different dimensions of emotional meaning. An evaluative map of exciting places differs from a map of relaxing ones, and these may differ for day and night. The grain of the image varies with the size of the area evaluated, with a coarser-grained image for larger areas and a finer-grained image for smaller areas. These findings highlight the importance of considering the relevant context in conducting such evaluations to form policy.

The studies also show some shared patterns of response. At a broad level, the studies confirm the presence of shared meanings by inhabitants. Each study uncovered areas of agreement on prominent places and meanings associated with them. When studies examined the features underlying these meanings, they also suggested some common themes. Likability tends to be associated with features related to order and social meanings (in particular, perceived social status). These features include nature, openness, upkeep, and history. When studies examined dimensions of emotion, they confirmed the relevance of pleasantness, arousal, relaxingness, and excitement for emotional appraisals of urban settings. They also show different meanings associated with different areas and sociocultural and environmental features. Taken together, the studies demonstrate various approaches, but they confirm that planners can use research on evaluative meanings to derive consensual directions for policy on community appearance. Several of the studies—Knoxville, Gilbert, Columbus—moved from research to policy. Campus planners have also used evaluative maps of a small campus to help create a campus master plan (Greene & Connelly, 1988; Shoen, 1991).

Notes

1. $F = 7.158$, 2 df, $p < 0.01$, $R^2 = .384$.

2. Because the same respondents did both tasks, the responses to the first task (selecting districts) may have influenced the subsequent evaluations.

3. These two factors account for 96% of the variance.

4. These two factors accounted for most of the variance (76.8%) and also fit the five clusters of wards.

5. Nighttime scenes score as significantly less exciting (t = 2.56, 19 df, $p < .05$), less interesting (t = 4.486, 19 df, $p < .001$), and marginally less pleasant (t = 1.98, 19 df $p = .06$), with mean differences of .22, .31, and .26 respectively on the 5-point scales. Day and night scores for relaxingness and fear do not differ significantly.

6. Pearson $r = .768$, $p < .001$; Spearman $r = .635$, $p < .01$.

7. The analysis used a procedure called canonical correlation. It created groupings of the evaluative ratings that related to one another and groupings of the physical characteristics that related to one another, and it simultaneously related the grouped evaluative scores to the grouped physical characteristics.

8. The group labeled *Natural and Openness* consisted of four scales: naturalness, openness, complexity, and absence of vehicles. The group labeled *Preference* consisted of six scales: pleasant, relaxing, interesting, exciting, safe, and active.

9. The group labeled *Disorder* consisted of nine scales: complexity; illegibility; incoherence; familiarity; nonuniform lighting; nuisance elements—wires, poles, fences, trash cans; deterioration; mystery; and vehicles. The group labeled *Arousingness* consisted of five scales: exciting, active, fearful, distressing, and interesting.

10. *Legibility* consisted of the scales vehicles, legibility, and absence of mystery.

11. *Visibility* consisted of three scales: brightness, naturalness, and the absence of mystery.

12. *Built Nuisance Elements* consisted of three scales: prominence of nuisance elements, brightness, and absence of naturalness. *Distress* consisted of two scales: distressing and unpleasant.

13. *Naturalness and Openness* consisted of naturalness, openness, complexity, mystery, and absence of vehicles. *Preference* consisted of pleasant, relaxing, interesting, and exciting.

14. *Disorder* consisted of complexity, illegibility, incoherence, nonuniform lighting, nuisances, vehicles, and atypicality. Arousal consisted of exciting, active, fearful, and distressing.

15. *Visibility* consisted of brightness, legibility, uniform lighting, and complexity. *Safety* consisted of safe, interesting, and active.

7

Shaping the Evaluative Image

The public agrees on its visual likes and dislikes, and evaluative maps reveal that agreement. By meeting public values, improvements in the evaluative image can make places more compatible to human purposes. They may also affect group identity, health, and well-being (Perkins, Meeks, & Taylor, 1992; Perkins, Wandersman, Rich, & Taylor, 1993; Taylor, 1989; Ulrich, 1983; Wilson & Kelling, 1982). Communities, assuming a separation between appearance and function, may leave decisions on appearance to designer intuition. But appearance can affect function. Our image, or our subjective knowledge of facts and values, influences behavior. If we value one place as better or more appealing than others, we will more likely want to go to it. Different appearances convey different meanings about activities (Cherulnik, 1991; Cherulnik & Wilderman, 1986; Nasar, 1989b; Sadalla, Verschure, & Burroughs, 1987), and different values may apply to different purposes. For example, people may value excitement for entertainment purposes, such as those in Times Square, but not for therapeutic purposes. For the latter, they may value a calming ambiance. Some residents may value a high-status appearance. Others may not. For certain public places, communities may value a friendly and welcoming ambiance. The visual character can affect these meanings and other behavioral functions (Figure 7.1).

Different places have different purposes, which in turn require different environmental characteristics (Brill, 1970; Sanoff, 1977, 1989).

Figure 7.1. Appearance Conveys Meaning About the Type and Quality of Activity
Photography by Jack. L. Nasar.

Some places, such as a neighborhood park, may serve their intended purpose better if they are designed to encourage interaction. Others, such as places for contemplation, may work better if they are designed to discourage interaction. Other characteristics, such as the level of familiarity, privacy, legibility, formality, and adaptability, may also affect the functioning of places. Achievement of the desired function depends on the ambiance.

To learn about how the design of visual features affects various psychological and social functions such as the ones discussed above, one could empirically study the relationship between specific visual features and a psychological or social objective. For example, a set of studies of lighting found that specific characteristics of lighting affected people's perceptions of the privacy, clarity (legibility), or pleasantness of a place (Flynn, 1988). As an example of how a combination of visual features may affect functioning, consider Paley Park, a popular New York gathering place, described by William Whyte (1980) in his research on livable urban spaces. The park attracts intense use. It has a formal entrance distinguishing it from the street and its neighbors, but, due to its movable chairs and activity, it looks informal and adaptable inside. Its organizational structure—defined entrance, food vendor by the entrance, sitting area, waterfall in the back—and simple and defined shape give it legibility. It also offers privacy through the waterfall, a nonvisual feature that buffers the ambient sound. Though Paley Park works for many reasons, the fit of the ambiance to desired activities adds to its desirability.

Some Generic Appearance Guidelines

In developing a visual plan, communities can use guidelines gleaned from previous research (reviewed in Chapter 4, "The Elements of Urban Likability"). These guidelines set performance criteria that the designer should satisfy. The ways in which designers satisfy the guidelines can vary.

Suppose one wants to set guidelines for creating a *pleasant* appearance. The research-based guidelines, or visual quality program, might call for the following:

1. Natural elements (vegetation and water) dominating built elements
2. Moderate complexity
3. High coherence, legibility, and compatibility of parts
4. Panoramas and defined open space
5. Easy maintenance
6. Styles that look historical

The research suggests that guidelines for an *exciting* appearance would differ. They might call for the following:

1. Built elements to dominate
2. High complexity
3. Low coherence and compatibility, bright colors, and high contrast between elements
4. High levels of movement and activity

The guidelines for a *relaxing* appearance might call for the following:

1. Natural elements to dominate
2. High coherence
3. Built elements, where they occur, to look
   a. Fitting or compatible with one another and nature
   b. Familiar and historical

Guidelines for a *high-status* appearance might call for the following:

1. Large open space
2. Well-kept and organized natural elements
3. Large, free-standing Tudor and Colonial-style buildings
4. Ornamentation

Some of the guidelines remain vague. We need additional research to identify in more concrete terms the way to achieve the desired end. Consider, for example, coherence. How does a plan achieve coherence and compatibility? How does a design review panel evaluate it in a design? For signs, coherence relates to decreased contrast—reduced sizes, brightness of colors, and relative contrast of colors (Nasar, 1987). For buildings in natural settings, compatibility or fittingness relates to building size, use, and the character of the site (Wohlwill, 1979). Building-to-building compatibility relates to redundancy of surface elements (Groat, 1984). Still, we need research identifying other factors that make a place appear coherent or compatible. Similarly, we need additional research on the components that contribute to an appearance as historical, defined open space, or moderate complexity. Even for naturalness, research can identify the organization and character of elements desired.

Using the Method for Design Policy

Communities may correctly question whether the guidelines will apply to their environmental and sociocultural context. Though research repeatedly shows consensus on evaluative response, unique requirements may apply to certain places, subcultures, or purposes. As an alternative to basing a visual plan on previous research, this book shows examples of research to derive appearance guidelines for the particular situation.

Evaluative maps provide information useful for planning the future appearance of a city or area. They give a handy snapshot of the perceived quality of the prominent elements in the city; they put the information in a form easily communicated to the public and decision makers; and they point to recommendations and controls for visual quality. Some findings—such as people's preference for vegetation and dislike of billboards—seem obvious. But by demonstrating widespread and consistent support for certain objectives, evaluative maps can help change those obvious preferences into physical realities. The approach fits the consumer orientation of legislation (Zube, 1980, p. 7) and U.S. Supreme Court decisions that place a higher burden on communities for their public regulatory actions (see, for example, *Daubert v. Merrell Dow,* 1993; *Dolan v. City of Tigard,* 1994; *Nollan v. California Coastal Commission,* 1987). It can give communities a nonarbitrary basis for guidelines relating specific physical requirements to the perceived appearance. The approach can also serve as a way to involve people in decisions that affect them, and it provides elected officials safe (popular) actions to take. Legislators, knowing the importance of responsiveness to their public constituencies, can appreciate the value of a public survey, especially if they understand its objectivity, reliability, and limitations (Zube, 1980, p. 37).

In Knoxville, where a quasi-public body sponsored the research, the results prompted several improvements. The city launched efforts to remove over 40 billboards from highways. In the central business district, a public-private partnership installed new unifying paving, vegetation, and barriers to hide autos. Inner-city neighborhoods held clean-up, fix-up campaigns. The city extended the well-liked Dogwood Trails.

In the university area, the study of the commercial signscape resulted in an overlay graphics code for the area. The city council approved a special set of sign regulations that applied just to the university area commercial strip. Presentation of the research at various community forums built community support for the changes. This led to a task force composed of residents, businesspersons, and city staff that revised the code based on the study recommendations. Specifically, the study results pointed to a need for a reduction in the contrast and obtrusiveness of the signs and for the maintenance of a moderate level of diversity. To reduce contrast and obtrusiveness, the code called for additional restrictions on the size, lighting, and brightness of colors of signs and the elimination of all free-standing signs and projecting signs. To maintain some diversity, the code allows some variability in height, shape, and colors. The survey results drew the full support of the residents and business community, which agreed with the findings and saw the favorable effect of the preferred sign character. When local planners brought the recommendations with the full support of the community to the city council, the council approved the changes. Thus, not only do surveys identify design directions, they also help shape recommendations into policy.

Cities can use a variety of regulatory, financial, administrative (Shirvani, 1985), and persuasion techniques to shape urban form. They can use codes for zoning, land use, buildings, and street graphics to control the location and character of development. The codes may include criteria for appearance. They can use financial mechanisms such as low-interest revolving loans and tax rebates to entice people into redeveloping in line with appearance guidelines. Cities can use other incentives, such as public-private partnerships or the offer of extra space for designs that provide certain visual amenities. For example, a city may add vegetation if a private party agrees to maintain it. A city may agree to give each business on a strip an awning or a sign for coherence if the business removes its nonconforming sign. Communities can also use design review to control appearance. Design review often relies on the personal discretion of the reviewers because of the absence, vagueness, or contradictory nature of design criteria, however. Evaluative image findings can help make the criteria more precise, thus making the development process more efficient for everyone involved.

In one resourceful administrative procedure, Annette Anderson, director of the Knoxville Neighborhood Design Assistance Center, noticed that many public and private agencies have access to public rights of way. This means that shortly after the city improves a street or sidewalk, the phone company, gas company, or some other agency could tear up the same site to perform its own work. Anderson created a form for each agency to complete so she could coordinate projects. Though participation was voluntary, she soon discovered that people at the agencies believed they had to complete the form to get approval for a projcct. Her simple administrative procedure coordinated development and removed some signs of decay from public sidewalks and streets.

Communities can make information available to relevant groups. For example, our public presentation of the findings on public reactions to different signscapes led one developer to use more restrictive guidelines before the city adopted them in the revised code. Communities can also give public notice and awards to desirable development, as many communities do when they give design awards. San Diego has an Orchids and Onions Program, in which the public votes on "orchids" (design contributions "that bring a sweet refreshing fragrance") and "onions" (design disappointments that "bring tears to your eyes"; Anthony, 1991, p. 150). A jury selects the finalists and makes the announcements at a large public banquet. These approaches bring people's attention to desired forms in a positive way. Such positive techniques, reinforcing desired behaviors rather than punishing undesirable ones, will more likely succeed in shaping behavior in the desired direction (Cone & Hayes, 1980). This suggests the value of offering incentives for desirable development rather than trying to punish undesirable development through code enforcement.

Beyond research for policy, I see some future directions and questions for research on evaluative meaning. Regrettably, I did not do a follow-up study to test the effects of the changes in Knoxville and Chattanooga on the evaluative image, but one use of evaluative maps involves evaluating the effect of changes after implementation. Evaluative maps can help chart the change in city appearance over time. Is it improving or getting worse? What areas are improving or getting worse? For example, a study can compare the proportion of favorable responses before and after a change to see if the image improved. It can compare the evaluations of different areas before and after a change to assess the effect of the change on the evaluative image to residents and nonresidents. Many cities undertake extensive downtown renovation projects. By assessing the evaluative image before the project, they can identify some directions for change. By assessing the evaluative image after the project, they can see the degree to which it achieved the intended results and the degree to which they need additional changes. This allows them to assess the efficacy of the policy to "modify, fine tune or abandon old policies or develop new policies" (Zube, 1980, p. 1). Such ongoing analysis of the public evaluative image—identifying visual strengths to preserve and visual problems to remove or change and identifying the qualities of these elements—can serve as the basis for planning the future visual form of a community.

Future Directions for Design and Research

Studies show different dimensions of emotional and social meaning, such as preference, excitement, distress, arousal, friendliness, and status. Preference may stand as an aggregate of these meanings, but if satisfaction with a city has to do with its compatibility to purposes, then different kinds of places should probably have different kinds of meanings and associated environmental characteristics. To shape cities for these meanings, we need additional research on the physical and social influences on the different dimensions of meaning.

Evaluative maps can also show similarities and differences among cities. Because experience plays a role in the evaluative image and meaning, we need to consider potential variations in the image related to experience. Such information can broaden our knowledge of ways to enhance the city image. We need studies of evaluative images and associated physical-sociocultural characteristics for a wider range of cities. In what ways do factors such as size, age, climate, location, and forces that influence city growth affect the evaluative image? We might also look at the evaluative images of popular cities, such as San Francisco, to identify the keys to their success and to examine the transferability of those concepts elsewhere. This book shows examples of evaluations for different scales of environment and the kind of information obtained. We need further work on evaluative images at different scales and on the appropriateness of different methods for various scales.

In addition to looking at different places and scales of place, we need to look at the responses of different groups. Because planning takes place around the world, we should consider the extent to which the findings for the United States apply elsewhere. Do the findings work in South America, Europe, the Middle East, or Asia? Cities also have many kinds of inhabitants. To maintain and enhance the quality of the environment for them, we need to look at the evaluative images of different groups and subcultures. We should also consider the evaluative images of special populations, such as children, the disabled, the elderly, transit riders, and others who, through their special environmental experience, may have unique evaluative images. At a more general level, we need to look at the relation of certain sociocultural characteristics to the evaluative image. Lifestyle, values, housing, and neighborhood choices relate to population characteristics such as social class, ethnicity, family structure, and stage in life cycle (Brower, 1996; Michelson, 1976). Evaluative meanings and the features associated with the meanings may also vary with such sociocultural characteristics. Such information has practical value. It would allow planners to make preliminary inferences about the needs of a neighborhood from census data instead of using more costly and time-consuming surveys. Planners could check their inferences at public meetings about the plans. With a detailed knowledge of the evaluative image for different place and population characteristics, communities can better plan solutions to satisfy everyone. Until then, communities can rely on knowledge of the shared public image.

Finally, we need additional study of the ways in which evaluative meaning changes over time. This requires further study of the evaluative image related to the season, day of the week, and time of day or night. Such studies can help refine the method, the assessment, and the application of evaluative images. Because we inhabit our communities over time, successful designs must respond to the variations in experience over time. If investigators know the conditions for which the evaluative image will likely vary, they can include those conditions in the study. When the features associated with a positive image at one time evoke a negative image at another time, we may need a new solution that works in both conditions. For example, diffuse lighting and permeable vegetation can make deflected vistas appear more open after dark, when these otherwise desirable features may become fearful.

This book set out to develop methods and ideas for improving the appearance of places. By examining various approaches and findings, it suggests some directions for research. One key direction for research involves the application and subsequent evaluation of the use of the evaluative image in urban design practice. Beyond that, important directions for research have to do with disaggregating the image. This means looking at different aspects of meaning, looking at different cities in the United States and elsewhere, looking at variations in scale and the methods appropriate to different scales, considering sociocultural group differences, and considering changes in the image related to time.

Studying the evaluative image has policy implications. By tapping public opinion and meanings, such study should give accurate predictions of future public reactions. It can serve in a variety of public and private actions, such as information dissemination, design review, incentives, public and private partnerships, zoning, codes, and enforcement. As part of a comprehensive plan or urban design, information about the evaluative image can help make changes in visual form more relevant to that public. Armed with information about popular preferences and meanings, what affects them, and how they change and develop, city designers can guide the image of the city toward becoming more likable, meaningful, and livable.

Appendix: Visual Quality Programming

Different methods can be used to uncover directions for community appearance. Each method yields a particular kind of information and has its particular set of potential biases. The selection of an appropriate method involves several decisions. You need to consider the fit of the method to the research purpose. Some research explores what is going on in relation to a particular phenomenon. Rather than specifying hypotheses in advance, this kind of research attempts to discover the hypotheses. For this purpose, you might get public opinion (in response to maps or in verbal response) about liked areas and disliked areas and the reasons for the evaluations. You might have people rate districts by name or as seen on a map. You might obtain ratings of various scenes in an area. You could have people build or manipulate scale models of an area (Nelessen, 1994). You could ask them to give examples or images of places that have the kind of appearance they want and ask them the reasons for their choices. You could have them describe and evaluate a preselected set of images, not necessarily from the area in question.

Other research attempts to demonstrate a hypothesis or to refute competing or alternative hypotheses. Controlled experiments fit these purposes best. For them, you might use scenes systematically manipulating the appearance variable of interest or selected for variation on that variable. For example, to test the effect of openness, you might systematically manipulate the amount of openness in a scene, or you might select scenes that vary in openness. Make sure that you control other variables that might affect response. You could also use nonexperimental designs. For this, you might obtain responses to various scenes neither manipulated nor selected for variables of interest. Then use various statistical techniques to examine relationships between variables. Research can also replicate other studies to test the stability of the original findings. For this purpose, you might use the same methods as the earlier study, or you might vary them. Whatever your research purpose, each method has potential shortcomings and biases. The presence of such biases points to the use of multiple methods and analyses as a way to get a better understanding of likely cause.

In addition to considering the research purpose, the investigator also faces a trade-off between practical and scientific concerns. A community may want a study within limited costs, resources, and time. These represent practical concerns. The community may want the results to apply to the population, settings, and measure of interest. This involves *external validity.* The community may want the results to eliminate rival hypotheses about causal relationships between the variables examined. This involves *internal validity.* How can you achieve practicality, external validity, and internal validity? Four methodological choices (Craik, 1970) can affect these outcomes:

1. Form of data collection
2. Selection of observers
3. Presentation of the environment
4. Assessing the environment

Let's consider each of the methodological choices in more detail.

Form of Data Collection

You can gather your data through mailed questionnaires, telephone interviews, personal interviews, or some combination of these methods (Judd, Smith, & Kidder, 1991). Mail surveys tend to cost the least, but they suffer from low response rates. Even with multiple letters and postcards, response rates may only reach 70% (Dillman, 1978). Results from a study with a low response rate may not apply to the population because you do not know if the respondents differ systematically from those who chose not to respond. Mail surveys also eliminate persons who cannot read, and you lose control of which person in the household completes the survey.

Telephone interviews fall in the middle on cost. They can get a fairly high response rate and, with random digit dialing, a representative sample. Phone interviews represent an efficient way to get a high-quality sample and to probe for details. They allow you to select the person in each household you want to interview. They give you control over the order of questions. For research on visual quality, however, phone interviews may have a shortcoming: You cannot use visual aids such as photographs, maps, or models.

In-person interviews tend to cost the most, but they also tend to get a high response rate and a representative sample. As with phone interviews, in-person interviews allow you to probe for detail, control question order, and select the household member to interview. In addition, they allow the use of visual aids. In-person interviews may introduce some interviewer bias, however. The interviewer may unknowingly sway respondents' answers so the answers do not accurately reflect respondents' real feelings. Research shows that various factors may lead respondents away from their natural response. Some of these factors include interviewer expectations (Rosenthal, 1976), the desire of respondents to give socially desirable answers (Rosenberg, 1965), and the desire of respondents to figure out the research so they can appear normal in their answers (Orne, 1962). Though research design can reduce these biases by keeping interviewers blind to the research purpose, it may not eliminate them.

As an alternative, the interviewer can leave surveys with people and offer to retrieve them later. As with mail surveys, this drop-and-retrieve approach depends on the respondent's ability to read, and this may eliminate some people from the sample. Otherwise, the approach has many benefits. It can get a high response rate and a representative sample. It tends to have a relatively low cost. It can include visual aids. Respondents can answer in anonymity, leading them to feel more comfortable answering honestly.

Thus, two methods stand out as efficient ways to get good information—telephone interviews and drop-and-retrieve surveys. The first does not allow the use of visual aids, but it does allow the interviewer to probe. The second reduces the interviewer's ability to probe for details, but it allows for visual aids. Used together, the strengths of each one can make up for the shortcomings of the other.

Selection of Observers

Whom and how many people do you study? You must identify the relevant user population and decide how to select individuals from this population. The sample should come from groups and individuals likely to experience the place evaluated. This might include residents, passers-by, regular visitors, occasional visitors, and others, or this might include groups similar to future users.

For a relatively small population, such as residents on one block, you might simply try to get responses from everyone. Otherwise, you must use some sampling procedure to select respondents from the population: *probability sampling* or *nonprobability sampling.* The first allows you to estimate the extent to which results apply to the population. The second does not, because the probability of any individual or group of individuals getting into the sample remains uncertain.

For external validity, you should seek a geographically representative sample of all residents of the area. Familiarity and preference for places in a city or neighborhood may depend on the observer's place of residence (Gould & White, 1974). Frequency of use, which relates to distance from the observer, also strongly influences people's judgments of the meaningfulness of places (Steinitz, 1968). You should also seek a sample representative of the diversity of people in the area. *Probability sampling* can get the desired sample. Probability samples include *simple random samples, stratified samples,* and *cluster samples.* To get a *simple random sample,* you need a complete list of persons in the population of interest from which you select names at random. For a *stratified random sample,* you split the population into relatively homogenous groups (such as socioeconomic groups, where responses across groups may vary). You might deliberately target subgroups so you can test the stability of the evaluative image across these groups. Then you draw a random sample from each group. Compiling the list, selecting names, and visiting places for interviews require considerable time, money, and resources. For

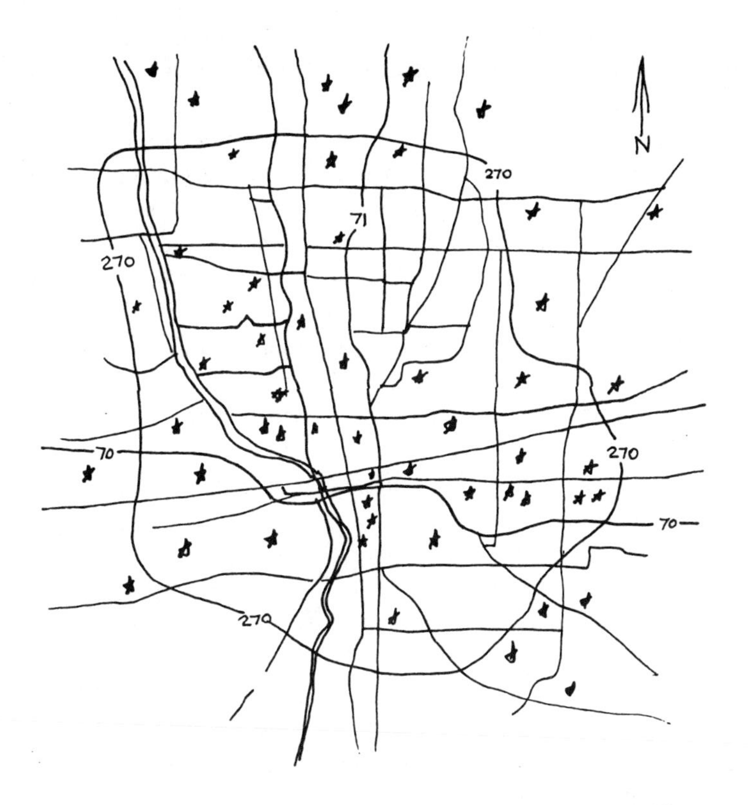

Figure A.1. Probability Sampling Leads to Geographic Diversity and Allows an Estimate of Applicability of Results to the Population
Drawing by Jack L. Nasar.

phone interviews, however, these approaches become more efficient because the interviewer need not visit the geographically diverse set of respondents (Figure A.1).

For in-person interviews, *cluster sampling* can save time. You first select clusters, such as neighborhoods or census tracts, on the basis of location. Then you sample within them at smaller and smaller levels. For example, from a full list of city census tracts, you could first select a random set of census tracts. For each selected tract, you could randomly select a census block, and within each block you could randomly select residents to interview. With some potential loss in external validity, cluster sampling can make in-person interviews more workable by narrowing the areas for interviews.

Compared with probability sampling, *nonprobability* sampling (or *opportunity* sampling) is more efficient, but it may have a cost. It will likely sacrifice the ability to estimate how well the results apply to the population. An opportunity sample involves selecting readily available respondents, such as passersby on a street. Because interviewers may unknowing select a certain kind of person to interview, thus introducing a bias, the interviewer should set up some standardized procedure for selecting respondents in advance. For example, the interviewer might decide to interview every fifth passerby of the opposite sex of the previous interview. For users of a park, the interviewer might randomly select seats to approach for interviews. To get a broader sample, the interviewer might attempt to vary the day and time of the interviews. For example, interviews might take place on weekdays and weekends in the early morning, midmorning, early afternoon, late afternoon, early evening, and late evening. The interviewer might also try to select public places that cover a broad sociodemographic spectrum of the population. For example, you could choose different kinds of public places, such as shopping centers, parks, and theaters, in different census tracts throughout the city. You could select tracts for their geographical diversity and for their varied population characteristics that represent the social classes, life cycle, and other characteristics of the city population. To run comparisons across groups (such as males and females), the sampling plan must capture an adequate number of people in each group. For population estimates, you can weigh group scores to reflect their actual proportions in the population. Though nonprobability samples may get skewed results, they can yield a sample representative of certain populations of interest. For example, you could get a sample representative of pedestrians passing by an area of interest such as a waterfront or a site for a new building. The pedestrians may represent the population to which you want to generalize: the nonprobability sample gets you the relevant respondents.

You also have to decide on the number of people to interview. In many public opinion polls, the pollster may report the percentage of a population in favor of an issue or candidate, the margin of error, and the probability of getting that error. The issue of whether the percentage of response to a particular item applies to a broad population usually requires a relatively large sample. Books of statistical tables usually show tables of sample sizes for various population sizes, margins of error, and confidence levels. For a city, you may need a sample of more than

1,000 respondents. Though such a sample is desirable, urban design decisions may not need the precision that such a sample gets. Smaller samples can identify the desired features.

When you want to make comparisons across groups or across conditions, the decision on sample size differs from the decision on representativeness of the sample. For comparisons, you need a sample large enough for valid statistical comparisons between experimental conditions. A sample of at least 30 persons per condition will meet the statistical assumptions for the parametric test of differences between groups, and this size sample can get the desired effect (Stamps, 1992a).

Some experimental designs have the same people respond to each condition. Researchers refer to this as a *within subject* or *repeated measure design* because the analysis compares the responses of the same persons to the different conditions. For example, in our study of the evaluative image of a signscape, we had each respondent respond to nine different signscapes systematically manipulated for complexity and coherence. The nine signscapes represented our experimental conditions. For a repeated measure design, the analysis needs a sample of only 30 persons.

Another research design compares responses of different respondents across conditions. Researchers refer to this as a *between subject design,* reflecting the comparison between different respondents. For example, in our signscape study, we compared responses of different groups, such as males and females. The groups represented between-subject conditions. We could have also had different groups assigned at random to each of the nine signscape conditions, in which case the nine signscapes would represent between-subject conditions. To compare responses of these kinds, each condition must have at least 30 members. The comparison of males versus females would have required a sample of 60 (30 males and 30 females), and the between-subject comparison among the nine conditions would have required a sample of 270 (30 persons for each of the nine signscapes).

Presentation of the Environment

You must choose an appropriate mode for presentation of scenes and for sampling scenes. Direct exposure to the actual environment seems the

TABLE A.1 Veridicality and Experimental Control of Some Media of Presentation

Similarity to Actual Experience	*Presentation Media*	*Experimental Control*
More Realistic	Direct on-site exposure to a place	Less control
↓	Color video or film of the place	↓
	Color photos or slides of the place	
	Color photos of model	
Less Realistic	Drawings and models	More control

Note: Black-and-white images are less realistic. A photomontage or computer manipulation of photo, film, or video can yield realistic and controlled images.

best for application. It comes closest to the ordinary environmental experience. Taking many people to various sites is inefficient, however, and the numerous extraneous factors on-site can interfere with experimental control, making it difficult to establish cause. To improve efficiency and experimental control, researchers use some form of simulation of the environment—recalled places, photography, photomontages, computer simulation, video or film, drawings, or models. These options vary in the degree to which they approximate direct on-site experience and offer experimental control (see Table A.1).

Several studies discussed in this book relied on respondents' recall of places they had experienced. The studies of Knoxville, Chattanooga, and German Village relied on respondent recall for parts of the city; the studies of Tokyo, Vancouver, Paris, and the university neighborhood obtained recalled impressions of selected districts. In each case, the scene selection came from the observers' recall. The researchers did not select or show the observers scenes or representations of scenes. The observers defined the scenes and variables of relevance to them. This approach is convenient, and it also has likely applicability to actual recall of familiar environments (Herzog, Kaplan, & Kaplan, 1976). Does this approach get at people's direct perception of places? Several studies have confirmed the reliability and validity of sketch maps and recall tasks (Baird, 1979; Evans, 1980; Howard, Chase, & Rothman, 1973; MacKay, 1976).

Nevertheless, the evaluative image of a city may work like a nested hierarchy, with the city image a collective result of neighborhoods and neighborhood images the collective results of blocks. Planners should not expect a study of the evaluative image of a full city to capture the details present in the image of a neighborhood. Nor should they expect a study at the neighborhood scale to capture the details present in the evaluative image of a block. Getting at these finer-grained experiences might require separate studies of smaller-scale places, or studies using other modes of presentation, such as photography.

Responses to *photography* accurately reflect on-site experience; and responses to color photos or slides more closely reflect on-site response than responses to other simulations such as drawings, models, or black-and-white photos (Feimer, 1984; Hershberger & Cass, 1974; Lyons, 1983; Nasar, 1988c; Oostendorp, 1978; Seaton & Collins, 1970; Shafer, Hamilton, & Smidt, 1969; Shafer & Richards, 1974; Zube, 1974). A review of over 152 environments evaluated by more than 2,400 observers (Stamps, 1990) found that preferences for places shown in photographs correlated highly with preferences on-site to the same places.[1] To control possible biases from extraneous variables, however, it makes sense to control the view, photographic angle, and other extraneous factors that may bias response. Airbrushing, photomontages, photos of models, or computer imaging also allow the simulation of different conditions. Research confirms that people do not detect the alterations via photomontage, and they give similar preferences to such altered images as they would experiencing the actual scene on-site (Stamps, 1992b). Research has also shown that scanned images, manipulated on a computer and printed as photographs, produce products indistinguishable from color slides and photos of real environments (Vining & Orland, 1989). People respond to them similarly. Computer imaging allows the researcher to sample a diversity of real scenes, digitize them, and alter them to control extraneous variables. This can create a more representative set of scenes with the desired realism at the same time that it allows for the systematic and controlled manipulation of physical forms.

Videotapes or *film* allow the addition of movement and visual sequences not present in still photos. You may approximate the feeling of movement by using a sequence of slides, though we need testing to evaluate the veridicality of this to actual movement. To gain additional control, you can videotape movement through scale models or use computer animation. *Drawings* allow for the manipulation of variables of interest while controlling others; *models* allow for three-dimensional manipulation, but the results may lack realism. Some research has found accurate perceptions of spatial behavior from scale models to actual environments (DeLong, 1977), though there are some contradictory findings. Other research has found no differences in recall for scenes presented as models or photographs (Dirks & Neisser, 1977). At present, we do not know whether these findings will generalize to evaluative appraisals obtained from models. Children may be less accurate with simulations than adults (Evans, 1980). The oblique perspective into a model differs from the eye-level view in real places; models may lack the level of detail present in the full-scale environments. Other research confirms potential problems with the accuracy of response to small-scale models (Evans, 1980). The Kaplans (1982) report success using eye-level photographs of the model to replicate the experience of actual environments. Responses to drawings or models may correlate to some degree with responses to the actual places (Stamps, 1992b), but they appear less accurate than responses to color photographs.

As for sampling scenes, you can choose from three broad approaches: *no investigator sampling, selective sampling,* or *representative sampling.* The use of no investigator sampling leaves the choice to the respondent, as in the recall studies of Knoxville and Chattanooga. It yields scenes of relevance to the observer's recall of the city. To bring this approach closer to direct perception, you might ask observers to take photos of places that have certain emotional meanings. *Selective sampling* involves choosing a variable of interest and then selecting scenes that vary on it. In using this approach, you should consider two potential issues. First, the scenes and variables selected may need to have relevance to ordinary experience for the results to generalize to ordinary environmental experience. Second, you may need to attend to the presence of other naturally occurring variables to improve your ability to identify cause. *Representative sampling* involves obtaining a variety of scenes representing the population of scenes. The sampling would follow procedures similar to those for getting a representative sample of a population. You might draw a random selection of scenes throughout a city. For example, social psychologist Stanley Milgram and his associates (Milgram et al., 1972) placed a 1,000 meter by 1000 meter grid over Manhattan and the four boroughs and selected a viewing point at each intersection, thinning out viewing points in the boroughs (Figure A.2).

You might draw a cluster sample of scenes. For example, a study of housing preferences selected residential areas from a zoning map at random. In each area, we selected streets, distance, and direction along

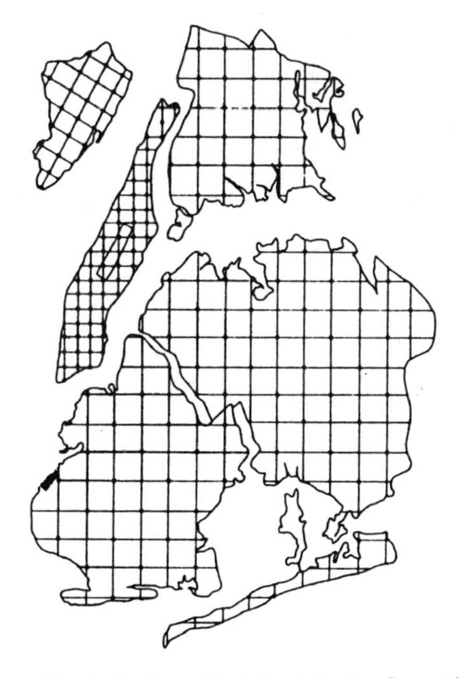

Figure A.2. Milgram Used a Grid as a Basis for Selecting Scenes in New York
SOURCE: Milgram et al. (1972).

the street and viewing angle at random (Nasar, 1988c). In each case, a random geographical sample of scenes might not represent the experienced city. Echoing the findings of Lynch (1960) and others, Milgram (Milgram et al., 1972) found differences in the accuracy of place recall related to centrality, to population flow and social or physical distinctiveness of places, and Steinitz (1968) found exposure (or visibility) of places to pedestrians as the most important determinant of people's knowledge about form and activity. To get at the city as experienced by the population, you might first develop a cognitive map establishing a hierarchy of places from the well-known to the unknown places. Then the sampling of places would randomly draw scenes from places, varying the number of scenes to reflect the relative imageability of the places. The sample would have more scenes from the imageable places than from the less imageable ones (Figure A.3).

Results from such a sample should apply well to the actual city experience. As with selective sampling, the presence of many extraneous

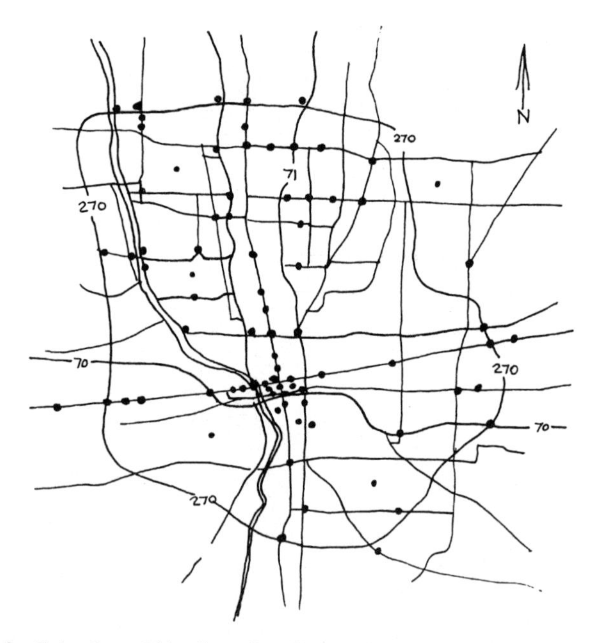

Figure A.3. Columbus, Ohio: Sampling Scenes Relative to Their Likely Imageability
Drawing by Jack L. Nasar.

variables hinders the identification of cause. Photomontage or computer image processing can regain some of that control, keep the sample representative, and keep the simulation realistic. Nevertheless, responses to a diversity of photos or to models can identify at a more general level the kinds of scenes that people like or dislike (Nelessen, 1994). If each scene represents its area as seen by the public, then the analysis can map responses to the scenes, yielding information of value for urban design.

Assessing the Environment

Humans have at least two kinds of responses to the environment—judgments of the physical characteristics of the environment and judgments of one's feelings about the environment. We call the first *perceptual-cognitive judgments* and the second *evaluative responses.* Although both represent subjective judgments made by observers, perceptual-cognitive judgments refer primarily to characteristics of the environment, and evaluative responses refer primarily to the observer's emotional response to the environment. To understand how people naturally respond to the noticeable features of the environment, the measures of physical characteristics of areas and emotional responses to them must reflect relevant dimensions to people.

Describing the Environment

Research findings have identified some noticeable features of the environment and salient aspects of evaluative response to various kinds of scenes (Kasmar, 1969; Nasar, 1988a; Oostendorp & Berlyne, 1978). For the environment, naturalness, openness, order, upkeep, historical significance, complexity, legibility, and mystery represent some key features that people notice in their surroundings. Other features may apply to specific contexts. To measure such features, you can either use direct physical measures or have observers rate or describe the environment. Direct physical measures make sense for descriptions of concrete features such as height, depth, openness, symmetry, or number of colors. Measures of more abstract features, such as the level of compatibility or the perceived prominence of naturalness, require judgmental measures. For example, a direct physical measure of the amount of nature in a scene may miss the perceived naturalness of a scene because the perception may depend on the arrangement as well as the amount of natural materials. A judgmental measure would capture the more relevant phenomenon: the perceived naturalness.

Evaluative Responses

For emotional response, prominent dimensions include pleasantness, arousal, excitement, calmness, and certain social meanings. Preference may yield a composite of the various evaluative and social meanings (Gould & White, 1974). In measuring emotional response, you should consider a potential limitation of verbal self-report. Such measures may identify "cold cognitions" lacking emotional involvement (Lazarus, 1984). To reduce this problem, you could first ask respondents whether they care at all about visual quality, and then proceed only if they say they do. You could also try to supplement verbal measures with physiological and behavioral measures. The former measure internal state (such as arousal), and the latter measure observable behavioral activity. Lacking such measures, you might get at internal state and behavior indirectly by asking about feelings of arousal or expected behavior in real situations. For example, we have asked people how much they would like to visit or spend time in a place, which of several houses they would take for free if they won it in a lottery, and which of several houses they would go to for help if they had a flat tire. A combination of verbal, physiological, and behavioral measures can enrich the picture of evaluative response.

Response Format

You also have to consider the response format for either perceptual-cognitive or evaluative responses. Response formats can vary from *open-ended* to *standardized response* (Table A.2). *Open-ended responses* do not constrain the respondent. Instead, they probe for the person's feelings, thoughts, and constructs. Whether written, spoken, or drawn, open-ended questions can gain rich information, but they may become difficult to quantify and sum across many observers. *Standardized responses* include various structured scales such as *adjective checklists,* sorting or *ranking methods,* and *semantic differential* scales. They lose some richness, but with relative ease, you can administer them to many people evaluating many places and quantify and tally the results. Interestingly, Gould and White (1974) report similar preference maps gleaned from open-ended responses and from preference ranking.

With an *adjective checklist,* you might have people check a list of adjectives that apply to the scene. This gets a variety of dimensions of responses to the scene and it yields *nominal* data. Similarly, having respondents identify places that they like or dislike yields nominal data. In each case, the data reveal categories but not the order or intensity of response in those categories. Sorting or *ranking* methods can obtain the order of scenes. They have respondents put places in order from the most to the least in the category, producing an *ordinal* scale. The *semantic differential* has people rate a scene on bipolar adjective scales such as pleasant–unpleasant or complex–simple. If they rate the scenes on 5-point (or 7-point) scales, the rating indicates the intensity of their response on an interval between 1 and 5 (or 7). It represents an *interval* scale. If they rate the intensity of their response on some interval between 0 and 100, this approximates a *ratio* scale.

TABLE A.2 Sample Open-Ended and Standardized Questions (Semantic Differential, Adjective Checklist, and Ranking Scales)

Open-Ended Questions

When you think of your city, what is the first thing that comes to mind?

Why did you think of that?

How do you feel about it?

What makes you think or feel that way?

What other things come to mind?

Semantic Differential

For each scale below, circle the number that best fits your feeling about the scene

Pleasant	1	2	3	4	5	6	7	Unpleasant
Boring	1	2	3	4	5	6	7	Exciting
High status	1	2	3	4	5	6	7	Low status
Calming	1	2	3	4	5	6	7	Distressing
Unsafe	1	2	3	4	5	6	7	Safe

Adjective Checklist

Check those adjectives that fit your feelings about the scene

____Pleasant	_____Safe	_____High status	_____Calming
____Low status	_____Distressing	_____Exciting	_____Arousing
____Boring	_____Unpleasant	_____Unsafe	_____Sleepy

Ranking Scales

For each question below, you can group scenes together that have a similar level.

Place the scenes in order from the one you like most to the one you like least.

Place scenes in order from the most boring to the most exciting.

Thinking of crime, place the scenes in order from the most safe to the most unsafe.

Place the scenes in order from the highest in status to the lowest in status.

The different kinds of scales require different kinds of statistical analyses, but statisticians have devised appropriate statistical analyses for each—parametric analyses for interval and ratio scales and nonparametric analyses for ordinal scales and frequencies in nominal categories. On the surface, it appears as if adjective checklists and ranking formats give up some of the precision possible from semantic differential scales. The former formats get only categories or ranks, whereas the latter can get intervals or ratios. Early research suggests that the type of measuring scale does not matter because each produces similar results (Gould & White, 1974). Subsequent analysis of eight different comparisons between scaling methods with more than 940 respondents confirms high correlations among the methods (Stamps, 1997). Ranking procedures have some additional practical benefits. They offer greater efficiency, allowing observers to respond to many scenes much more quickly than they could through the other formats; they offer a higher level of agreement among respondents than the other scales (Brush, 1976; Zube, Pitt, & Anderson, 1974); and, over a large number of respondents, data from rankings approximate the metrics of interval scales (Gould & White, 1974).

Whichever format you use, you need to consider some other issues in assessing perceptual-cognitive features or evaluative meanings. First, the order of presentation of scenes and the order of presentation of scales may bias responses. To lessen the potential effects, you should vary the order of both the scenes and the scales across observers. Second, respondents may exhibit *response set,* checking a point on each scale without carefully thinking about it. To lessen this bias, you should vary the positive and negative poles of the scales at random. Third, perceptual-cognitive and evaluative ratings may influence one another such that the analysis uncovers extraneous relationships rather actual relationships between environmental features and evaluations. For example, if the same person first rates the coherence of a scene and then rates its pleasantness, the first judgment might call attention to coherence, which in turn would unnaturally influence the pleasantness rating. If the person rates pleasantness first and then coherence, the first judgment might call attention to the evaluation, which could affect the judged coherence. Some perceptual-cognitive and evaluative terms may also have a semantic link, such that analysis would uncover a semantic relationship rather than one between the feature and evaluation. To lessen these biases, you should obtain independent ratings of the two kinds of variables: Have one set of observers judge features of the place and a separate set of observers evaluate the place. You should also work on the reliability and construct validity of the environmental descriptions. Where possible, use direct physical measures. If you use rating scales, take care to avoid tautological reasoning—using adjectives for perceptual-cognitive judgments and evaluations that say essentially the same thing. For example, the rated dirtiness and pleasantness of an area may show a relationship because the words have shared meanings. Try to define the scales clearly—describing the variable and the meaning of various points on the scale. You might use graphics to show the various points on the scale. Pretest each scale on scenes that vary on the scale and scenes that do not to ensure that the scale differentiates between the appropriate scenes. Pretrain judges on the use of the scales so they learn to use them consistently.

Various mixes in procedures for selecting observers, presenting the environment, getting evaluative response, and describing the environment apply to different situations. For city planning and design, the research may aim more for the practical application of results than for the testing of theory. Certainly, theory and the understanding that go beyond a particular application have importance, but ultimately urban design and planning try to apply the results. For purposes of application, the conditions of the research should approximate the real conditions while minimizing loss of control. One reasonable method to accomplish this involves interviewing a relatively small cluster sample for evaluative responses to maps or photos of sections of the city. You might select scenes for inclusion based on their visibility in the city. One group of respondents might rank the photographs of scenes for various evaluative scales (pleasantness, excitingness, distress, perceived status, friendliness, and desirability as a place to visit and spend time). Another group might produce different maps for each scale. Each group might also receive a checklist of sociocultural and physical features to explain the reasons for their responses. A separate phone survey could have a random sample of respondents give verbal evaluations along the same dimensions. A third group (perhaps planners) might judge various visual features of the scenes, and the investigator might measure various characteristics (such as openness, naturalness, and number of different elements). The statistical analyses could then uncover relationships between the evaluations and the place characteristics. Beyond answering immediate questions about visual quality, the appropriate methodological choices can become part of a database for future questions. This would allow communities to combine results from several studies to build a firmer knowledge base for guiding future evaluative quality.

Note

1. $r = .86, p < .001$.

References

Alexander, C. (1966). A city is not a tree. *Design, 206,* 46-55.

Anderson, W. S. (1991). *Children, image mapping and the general plan.* Gilbert, AZ: Gilbert Department of Planning.

Anthony, K. H. (1991). *Design juries on trial: The renaissance of the design studio.* New York: Van Nostrand Reinhold.

Appleton, J. (1975). *The experience of place.* London: Wiley.

Appleyard, D. (1970). Styles and methods of structuring a city. *Environment and Behavior, 2,* 131-156.

Appleyard, D. (1976). *Planning a pluralist city: Conflicting realities in Ciudad Guayana.* Cambridge: MIT Press.

Appleyard, D. (1981). *Livable streets.* Berkeley: University of California Press.

Appleyard, D. (1982). Three kinds of urban design practice. In A. Ferebee (Ed.), *Education for urban design* (pp. 122-126). Purchase, NY: Institute for Urban Design.

Appleyard, D., Lynch, K., & Myer, J. R. (1964). *The view from the road.* Cambridge: MIT Press.

Aragones, J. L., & Arredondo, J. M. (1985). Structure of urban cognitive maps. *Journal of Environmental Psychology, 5,* 197-212.

Baird, J. (1979). Studies of the cognitive representation of spatial relations. *Journal of Experimental Psychology: General, 108,* 90-106.

Beck, R. J., & Wood, D. (1976). Cognitive transformation of information from urban geographic fields to mental maps. *Environment and Behavior, 8,* 199-238.

Berlyne, D. E. (1971). *Aesthetics and psychobiology.* New York: Meredith.

Bishop, J. (1984). Passing in the night: Public and professional views of Milton Keynes. *Places, 1,* 9-16.

Blake, P. (1974). *Form follows fiasco.* Boston: Atlantic Monthly.

Bornstein, R. F. (1989). Exposure and affect. Overview and meta-analysis of research, 1968-1987. *Psychological Bulletin, 106,* 265-289.

Brill, M. (1970). *PAK: Planning aid kit.* Paper presented at the 123rd meeting of the American Psychiatric Association, San Francisco.

Brower, S. (1988). *Design in familiar places: What makes home environments look good.* New York: Praeger.

Brower, S. (1996). *Good neighborhoods, a study of in-town & suburban residential environments.* Westport, CT: Praeger.

Brown, B., & Altman, I. (1983). Territoriality, defensible space, and residential burglary: An environmental analysis. *Journal of Environmental Psychology, 3,* 203-220.

Brunswik, E. (1956). *Perception and representative design of psychological experiments.* Berkeley: University of California Press.

Brush, R. O. (1976). Perceived quality of scenic and recreational environments: Some methodological issues. In K. H. Craik & E. H. Zube (Eds.), *Perceiving environmental quality: Research and applications* (pp. 47-58). New York: Plenum.

Burgess, J. A. (1978). *Image and identity: A study of urban and regional perception with particular reference to Kingston-upon Hull.* Hull, UK: University of Hull.

Campbell, D. T., & Fiske, D. W. (1959). Convergent and discriminant validity by the multitrait-multimethod matrix. *Psychological Bulletin, 56,* 81-105.

Campbell, D. T., & Stanley, J. C. (1963). *Experimental and quasi-experimental designs for research.* Chicago: Rand McNally.

Canter, D. (1969). An intergroup comparison of connotative dimensions in architecture. *Environment and Behavior, 1,* 37-48.

Carp, F. M., Zawadski, R. T., & Shokron, H. (1976). Dimensions of urban environmental quality. *Environment and Behavior, 8,* 239-264.

Chadwin, M. L. (1975). The nature of legislative program evaluation. *Evaluation, 2,* 45-49.

Cherulnik, P. D. (1991). Reading restaurant facades: Environmental inference in finding the right place to eat. *Environment and Behavior, 22,* 150-170.

Cherulnik, P. D., & Wilderman, S. K. (1986). Symbols of status in urban neighborhoods: Contemporary perceptions of nineteenth-century Boston. *Environment and Behavior, 18,* 604-622.

Cone, J. D., & Hayes, S. C. (1980). *Environmental problems/behavioral solutions.* Monterey, CA: Brooks/Cole.

Cook, T. D., & Campbell, D. T. (1979). *Quasi-experimentation: Design and analysis issues for field settings.* Chicago: Rand McNally.

Cooper, C. (1972). Resident dissatisfaction in multifamily housing. In W. M. Smith (Ed.), *Behavior, design and policy aspects of human habitats* (pp. 119-146). Green Bay: University of Wisconsin Press.

Coren, S., & Girgus, J. S. (1980). Principles of perceptual organization and spatial distortion: The Gestalt illusions. *Journal of Experimental Psychology: Human Perception and Performance, 6,* 404-412.

Craik, K. H. (1970). Environmental psychology. In T. M. Newcomb (Ed.), *New directions in psychology* (Vol. 4, pp. 1-121). New York: Holt, Rinehart & Winston.

Craik, K. H. (1983). The psychology of the large scale environment. In N. R. Feimer & E. S. Geller (Eds.), *Environmental psychology: Directions and perspectives* (pp. 67-105). New York: Praeger.

Craik, K. H., & Appleyard, D. (1980). Streets of San Francisco: Brunswik's lens model applied to urban inference and assessment. *Journal of Social Issues, 36,* 72-85.

Craik, K. H., & Zube, E. H. (1976). *Perceiving environmental quality, research and applications.* New York: Plenum.

Daubert v. Merrell Dow. 509 U.S. 579, 113 S. Ct. 2787, 125 L. Ed. 2d 469 (1993).

deJonge, D. (1962). Images of urban areas. Their structure and psychological foundations. *Journal of the American Institute of Planners, 28,* 266-276.

DeLong, A. (1977). The accuracy of spatial perception by information in scale model environments. *Man-Environment Systems, 7,* 55-58.

Devlin, K., & Nasar, J. L. (1989). The beauty and the beast: Some preliminary comparisons of "high" versus "popular" residential architecture and public versus architect judgments of same. *Journal of Environmental Psychology, 9,* 333-344.

Dillman, D. A. (1978). *Mail and telephone surveys.* New York: John Wiley.

Dirks, J., & Neisser, U. (1977). Memory for objects in real scenes. The development of recognition and recall. *Journal of Experimental Child Psychology, 23,* 315-328.

Dolan v. City of Tigard. 512 U.S. 374, 114 S. Ct. 2309, 129 L. Ed. 2d 304 (1994).

Downs, R. M. (1976). Cognitive mapping and information processing: A commentary. In G. T. Moore & R. G. Colledge (Eds.), *Environmental knowing: Theories, research and methods* (pp. 67-70). Stroudsburg, PA: Dowden, Hutchinson and Ross.

Duncan, J. (1973). Landscape taste as a symbol of group identity: Westchester County Village. *Geographical Review, 63,* 334-355.

Duncan, J. S., Lindsey, S., & Buchan, R. (1985). Decoding a residence: Artifacts, social codes and the construction of the self. *Espaces et societes, 47,* 29-43.

Espe, H. (1981). Differences in perception of national socialist and classicist architecture. *Journal of Environmental Psychology, 1,* 33-42.

Evans, G. (1980). Environmental cognition. *Psychological Bulletin, 88,* 259-287.

Evans, G. W., Brennan, P. L., Skorpanich, M. A., & Held, D. (1984). Cognitive mapping and elderly adults: Verbal and locational memory for urban landmarks. *Journal of Gerontology, 39,* 452-457.

Evans, G. W., & Garling, T. (Eds.). (1991). *Environment, cognition, and action: An integrated approach.* New York: Oxford University Press.

Evans, G., Smith, C., & Pezdek, K. (1982). Cognitive maps and urban form. *Journal of the American Planning Association, 48,* 232-244.

Ewald, W. R., & Mandelker, D. R. (1977). *Street graphics: A concept and a system.* McLean, VA: Landscape Architecture Foundation.

Feather, N. T. (1964). Acceptance and rejection of arguments in relation to attitude strength, critical ability and intolerance of inconsistency. *Journal of Abnormal and Social Psychology, 69,* 127-136.

Fechner, G. T. (1876). *Vorschule der asthetik.* Leipzig: Breitopf & Hartel.

Feimer, N. (1984). Environmental perception: The effect of media, evaluative context and the observer sample. *Journal of Environmental Psychology, 4,* 61-80.

Fisher, B., & Nasar, J. L. (1992). Fear of crime in relation to three exterior site features: Prospect, refuge and escape. *Environment and Behavior, 24,* 35-65.

Flynn, J. E. (1988). Lighting-design decisions as interventions in human visual space. In J. L. Nasar (Ed.), *Environmental aesthetics: Theory, research and applications* (pp. 156-170). New York: Cambridge University Press.

Francaviglia, R. V. (1978). Xenia rebuilds: Effects of predisaster conditioning on post-disaster redevelopment. *Journal of the American Planning Association, 44,* 13-24.

Francescato, D., & Mebane, W. (1983). How citizens view two great cities: Milan and Rome. In R. M. Downs & D. Stea (Eds.), *Image and environment: Cognitive mapping and spatial behavior* (pp. 131-147). Chicago: Aldine.

Frankel, M. R., & Frankel, L. R. (1987). Fifty years of survey sampling in the U.S. *Public Opinion Quarterly, 51,* S127-S138.

A French view of New York: Perilous city to visit. (1972, January 24). *New York Times,* p. 1.

Gans, H. (1962). *The urban villagers: Group and class in the life of Italian Americans.* New York: Free Press.

Gans, H. (1974). *Popular culture and high culture: An analysis and evaluation of taste.* New York: Basic Books.

Gibson, J. J. (1979). *The ecological approach to visual perception.* Boston: Houghton Mifflin.

Gifford, R. (1980). Environmental dispositions and the evaluation of architectural interiors. *Journal of Research in Personality, 14,* 386-399.

Gleye, P. H. (1983). The breath of history: An investigation of urban features conveying historical identity in the townscape as observed in the postwar reconstruction of Munster, West Germany; with an application to Los Angeles. *Dissertation Abstracts International, 44*(05), 1606A. (University Microfilms No. AAI8321977).

Golledge, R. G. (1977). Multidimensional analysis in the study of environmental behavior and environmental design. In I. Altman & J. F. Wohlwill (Eds.), *Human behavior and environment* (Vol. 2, pp. 1-42). New York: Plenum.

Gould, P. (1973). On mental maps. In R. M Downs & D. Stea (Eds.), *Image and environment: Cognitive mapping and spatial behavior* (pp. 182-222). Chicago: Aldine.

Gould, P., & White, R. (1974). *Mental maps.* Middlesex, UK: Penguin.

Greene, T. C., & Connelly, C. M. (1988). Computer analysis of aesthetic districts. In D. Lawrence, R. Habe, A. Hacker, & D. Sherrod (Eds.), *People's needs/planet management: Paths to co-existence* (pp. 333-335). Washington, DC: Environmental Design Research Association.

Groat, L. (1982). Meaning in post-modern architecture: An examination using the multiple sorting task. *Journal of Environmental Psychology, 2,* 3-22.

Groat, L. (1983). *A study of the perception of contextual fit in architecture.* Paper presented at the 14th international conference of the Environmental Design Research Association, Lincoln, NE.

Groat, L. (1984, November). Public opinions of contextual fit. *Architecture,* pp. 72-75.

Groat, L. N., & Despres, C. (1990). The significance of architectural theory for environmental design research. In E. H. Zube & G. T. Moore (Eds.), *Advances in environment, behavior, and design* (Vol. 3, pp. 3-53). New York: Plenum.

Groves, R. M. (1987). Research on survey data quality. *Public Opinion Quarterly, 51,* S156-S172.

Gulick, J. (1963). Images of an Arab city. *Journal of the American Institute of Planners, 29,* 179-198.

Hanyu, K. (1993). The affective meaning of Tokyo: Verbal and non-verbal approaches. *Journal of Environmental Psychology, 13,* 161-172.

Hanyu, K. (1995). Visual properties and affective appraisals in residential areas. *Disseration Abstracts International, 56*(12), 4978A. (University Microfilms. No. AAI9612190).

Hardin, G. (1968). The tragedy of the commons. *Science, 162,* 1243-1246.

Harrison, J. D., & Howard, W. A. (1972). The role of meaning in the urban image. *Environment and Behavior, 4,* 389-411.

Harrison, J. D., & Sarre, P. (1975). Personal construct theory in the measurement of environmental images. *Environment and Behavior, 7,* 3-58.

Hart, R. (1979). *Children's experience of place.* New York: Irvington.

Hartig, T., Mang, M., & Evans, G. W. (1991). Restorative effects of natural environment experiences. *Environment and Behavior, 23,* 3-26.

Heise, D. R. (1970). The semantic differential and attitude research. In G. F. Summers (Ed.), *Attitude measurement* (pp. 235-253). Chicago: Rand McNally.

Hershberger, R. G. (1969). A study of meaning and architecture. In H. Sanoff & S. Cohn (Eds.), *EDRA 1: Proceedings of the first annual environmental design research association conference* (pp. 86-100). Raleigh: North Carolina State University.

Hershberger, R. G., & Cass, R. C. (1974). Predicting user responses to buildings. In G. Davis (Ed.), *Man environment interaction: Evaluations and applications, the state of the art in environmental design research—Field applications* (pp. 117-134). Milwaukee, WI: Environmental Design Research Association.

Herzog, T. R., Kaplan, S., & Kaplan, R. (1976). The prediction of preference for familiar urban place. *Environment and Behavior, 8,* 627-645.

Herzog, T., Kaplan, S., & Kaplan, R. (1982). The prediction of preference for unfamiliar urban places. *Population and Environment, 5,* 43-59.

Herzog, T., & Smith, G. A. (1988). Danger, mystery, and environmental preference. *Environment and Behavior, 20,* 320-344.

Hesselgren, S. (1975). *Man's perception of man-made environment.* Stroudsburg, PA: Dowden, Hutchinson and Ross.

Hinshaw, M., & Allot, K. (1972). Environmental preferences of future housing consumers. *Journal of the American Institute of Planners, 38*(2), 102-107.

Hogan, R. (1982). A socioanalytic theory of personality. In M. M. Page (Ed.), *Nebraska Symposia on Motivation* (Vol. 30). Lincoln: University of Nebraska Press.

Horayangkura, V. (1978). Semantic dimensional structures: A methodological approach. *Environment and Behavior, 10,* 555-584.

Howard, R. B., Chase, S. D., & Rothman, M. (1973). An analysis of four measures of cognitive maps. In W. F. E. Preiser (Ed.), *Environmental Design Research* (Vol. 1, pp. 254-264). Stroudsburg, PA: Dowden, Hutchinson and Ross.

Hull, R. B., & Revell, G. R. B. (1989). Cross-cultural comparison of landscape scenic beauty evaluations: A case study in Bali. *Journal of Environmental Psychology, 9,* 177-191.

Im, S-B. (1984). Visual preferences in enclosed urban spaces: An exploration of a scientific approach to environmental design. *Environment and Behavior, 16,* 235-262.

International City Management Association. (1984). *Facilitating economic development: Local government activities and organizational structures.* (Urban Data Service Report, 16, 11/12.) Washington, DC: International City Management Association.

Jones, B. (1990). *Neighborhood planning: A guide for cities and planners.* Chicago: Planners Press.

Judd, C. M., Smith, E. R., & Kidder, L. H. (1991). *Research methods in social relations* (6th ed.). Chicago: Holt, Rinehart & Winston.

Kang, J. (1990). Symbolic inferences and typicality in five taste cultures. *Dissertation Abstracts International, 51*(12), 3929A. (University Microfilms No. AAI9105142).

Kaplan, R., & Kaplan, S. (1989). *The experience of nature: A psychological perspective.* New York: Cambridge University Press.

Kaplan, S. (1975). An informal model for the prediction of preference. In E. H. Zube, J. G. Fabor, & R. O. Brush (Eds.), *Landscape assessment* (pp. 92-101). Stroudsburg, PA.: Dowden, Hutchinson and Ross.

Kaplan, S. (1995). The restorative benefits of nature: Towards an integrative framework. *Journal of Environmental Psychology, 15,* 169-182.

Kaplan, S., & Kaplan, R. (1982). *Cognition and environment: Functioning in an uncertain world.* New York: Praeger.

Kaplan, S., Kaplan, R., & Wendt, J. S. (1972). Rated preference and complexity for natural and urban visual material. *Perception and Psychophysics, 12,* 354-356.

Kaplan, S., & Talbot, J. F. (1983). Psychological benefits of a wilderness experience. In I. Altman & J. F. Wohlwill (Eds.), *Behavior in the natural environment* (pp. 163-201). New York: Plenum.

Kasmar, J. V. (1969). The development of a useable lexicon of environmental descriptors. *Environment and Behavior, 2,* 153-169.

Klein, H. (1967). The delimitation of the town-centre in the image of its citizens. In University of Amsterdam Sociographical Department (Eds.), *Urban core and inner city* (pp. 286-306). Leiden: E. J. Brill.

Knoxville/Knox County Metropolitan Planning Commission. (1988). *Knoxville area facts and figures.* Knoxville, TN: Author.

Koffka, K. (1935). *Principles of gestalt psychology.* Princeton, NJ: Princeton University Press.

Kuller, R. A. (1972). *A semantic model for describing perceived environment.* Stockholm: National Swedish Institute for Building Research.

Lang, J. (1987). *Creating architectural theory: The role of the behavioral sciences in environmental design.* New York: Van Nostrand Reinhold.

Lansing, J. B., Marans, R. W., & Zehner, R. B. (1970). *Planned residential environments.* Ann Arbor: University of Michigan, Survey Research Center Institute for Social Research.

Lazarus, R. S. (1984). On the primacy of cognition. *American Psychologist, 39,* 124-129.

Lee, L-S. (1982). Image of city hall. In P. Bart, A. Chen, & G. Francescato (Eds.), *Knowledge for Design. EDRA 13* (pp. 310-317). Washington, DC: Environmental Design Research Association.

Leff, H. L. (1978). *Experience, environment and human potential.* New York: Oxford University Press.

Leff, H. L., Gordon, L. R., & Ferguson, J. G. (1974). Cognitive set and environmental awareness. *Environment and Behavior, 6,* 395-447.

Lightner, B. (1993). *A survey of design review practice in local government.* Chicago: American Planning Association.

Livingood, J. W. (1981). *A history of Hamilton county.* Memphis, TN: Memphis State University Press.

Loftus, E. F., & Hoffman, H. G. (1989). Misinformation and memory: The creation of new memories. *Journal of Experimental Psychology: General, 118,* 100-104.

Lozano, E. (1974). Visual needs in the environment. *Town Planning, 43,* 351-374.

Lynch, K. (1960). *The image of the city.* Cambridge: MIT Press.

Lynch, K. (1976). *Managing the sense of region.* Cambridge: MIT Press.

Lynch, K. E., & Rivkin, M. (1959). A walk around the block. *Landscape, 8,* 24-34.

Lyons, E. (1983). Demographic correlates of landscape preference. *Environment and Behavior, 15,* 487-511.

MacKay, D. B. (1976). The effect of spatial stimuli on the estimation of cognitive maps. *Geographical Analysis, 8,* 439-452.

Magana, J. R. (1978). An empirical and interdisciplinary test of a theory of urban perception. *Dissertation Abstracts International 39*(03), 1460b. (University Microfilms No. AAI9105142).

Marans, R. W. (1976). Perceived quality of residential environments: Some methodological issues. In K. H. Craik & E. H. Zube (Eds.), *Perceiving environmental quality: Research and applications* (pp. 123-147). New York: Plenum.

Mars, P. R. (1996). Aesthetics as a survival mechanism: Towards a theory of architecture. *Dissertation Abstracts International, 57*(04), 1355A. University Microfilms No. AAI9625689).

Marsh, T. A. (1993a, October). *"Seeing Differently": Some observations and propositions regarding architects' and non-architects' perceptions of architecture.* Paper presented at the meeting of Crossing Boundaries of Practice Conference, Cincinnati, OH.

Marsh, T. A. (1993b). *Through others' eyes. An investigation of architects' and non-architects' perceptions of architecture.* Master's thesis, State University of New York, Buffalo.

Maurer, R., & Baxter, J. C. (1972). Images of the neighborhood and city among black, Anglo and Mexican-American children. *Environment and Behavior, 4,* 351-388.

Michelson, W. (1976). *Man and the urban environment.* Reading, MA: Addison-Wesley.

Michelson, W. (1987). Groups, aggregates, and the environment. In E. H. Zube & G. T. Moore (Eds.), *Advances in environment, behavior, and design* (Vol. 1, pp. 161-185). New York: Plenum.

Milgram, S., Greenwald, J., Kessler, S., McKenna, W., & Waters, J. (1972). A psychological map of New York City. *American Scientist, 60,* 194-200.

Milgram, S., & Jodelet, D., (1976). Psychological maps of Paris. In H. Proshansky, W. Ittleson, & L. Rivlin (Eds.), *Environmental psychology: People and their physical settings* (2nd ed., pp. 104-124). New York: Holt, Rinehart & Winston.

Moore, G. T. (1989). Environment and behavior research in North America: History, developments, and unresolved issues. In D. Stokols & I. Altman (Eds.), *Handbook of environmental psychology* (pp. 1359-1410). New York: John Wiley.

Moreland, R. L., & Zajonc, R. B. (1977). Is stimulus recognition a necessary condition for the occurrence of exposure effects? *Journal of Personality and Social Psychology, 35,* 191-199.

Moreland, R. L., & Zajonc, R. B. (1979). Exposure effects may not depend on stimulus recognition. *Journal of Personality and Social Psychology, 37,* 1085-1089.

Nasar, J. L. (1983). Adult viewers' preferences in residential scenes: A study of the relationship of environmental attributes to preference. *Environment and Behavior, 15,* 589-614.

Nasar, J. L. (1984). Visual preferences in urban street scenes: A cross-cultural comparison between Japan and the United States. *Journal of Cross-Cultural Psychology, 15,* 79-93.

Nasar, J. L. (1987). Effects of signscape complexity and coherence on the perceived visual quality of retail scenes. *Journal of the American Planning Association, 53,* 499-509.

Nasar, J. L. (1988a) Editor's introduction, Section D: Urban space. In J. L. Nasar (Ed.), *Environmental aesthetics: Theory, research, and applications* (pp. 257-259). New York: Cambridge University Press.

Nasar, J. L. (1988b). *Environmental aesthetics: Theory, research, and applications.* New York: Cambridge University Press.

Nasar, J. L. (1988c). Perception and evaluation of residential street-scenes. In J. L. Nasar (Ed.), *Environmental aesthetics: Theory, research, and applications* (pp. 275-289). New York: Cambridge University Press.

Nasar, J. L. (1989a). Perception, cognition, and evaluation of urban places. In I. Altman & E. H. Zube (Eds.), *Public places and spaces: Human behavior and environment* (Vol. 10, pp. 31-56). New York: Plenum.

Nasar, J. L. (1989b). Symbolic meanings of house styles. *Environment and Behavior, 21,* 235-257.

Nasar, J. L. (1990). The evaluative image of the city. *Journal of the American Planning Association, 56,* 41-53.

Nasar, J. L. (1994). Urban design aesthetics: The evaluative qualities of building exteriors. *Environment and Behavior, 26,* 377-401.

Nasar, J. L. (1998). *Design by competition.* New York: Cambridge University Press.

Nasar, J. L., & de Nivia, C. (1987). A post-occupancy evaluation for the design of a light pre-fabricated housing system for low income groups in Colombia. *Journal of Architectural and Planning Research, 4,* 199-211.

Nasar, J. L., & Fisher, B. (1993). "Hot spots" of fear of crime: A multiple-method investigation. *Journal of Environmental Psychology, 13,* 187-206.

Nasar, J. L., Fisher, B., & Grannis, M. (1993). Proximate cues to fear of crime. *Landscape and Urban Planning, 26,* 161-178.

Nasar, J. L., & Jones, K. (1997). Landscapes of fear and stress. *Environment and Behavior, 29,* 291-321.

Nasar, J. L., & Julian, D. (1985). Effects of labeled meaning on the affective quality of housing scenes. *Journal of Environmental Psychology, 5,* 335-344.

Nasar, J. L., Julian, D., Buchman, S., Humphreys, D., & Mrohaly, M. (1983). The emotional quality of scenes and observation points: A look at prospect and refuge. *Landscape Planning, 10,* 355-361.

Nasar, J. L., & Kang, J. (1989). A post-jury evaluation: The Ohio State University design competition for a center for the visual arts. *Environment and Behavior, 21,* 464-484.

Neilson, W. A., Knott, T. A., & Carhard, P. W. (Eds.). (1960). *Webster's new international dictionary of the English language* (2nd ed.). Springfield, MA: Merriam.

Neisser, U. (1976). *Cognition and reality.* San Francisco: Freeman.

Nelessen, A. C. (1994). *Visions for a new American dream.* Chicago: The American Planning Association.

Newman, O. (1972). *Defensible space. Crime prevention through urban design.* New York: Macmillan.

Nisbett, R. E., Krantz, D. H., Jepson, C., & Kunda, Z. (1983). The use of statistical heuristics in everyday inductive reasoning. *Psychological Review, 90,* 339-363.

Nollan v. California Coastal Commission. 483 U.S. 825 (1987).

Norberg-Schulz, C. (1965). *Intentions in architecture.* Cambridge: MIT Press.

Oostendorp, A. (1978). The identification and interpretation of dimensions underlying aesthetic behaviour in the daily urban environment. *Dissertation Abstracts International, 40*(2), 990B. (University Microfilms No. AAI053107).

Oostendorp, A., & Berlyne, D. E. (1978). Dimensions in the perception of architecture: Measures of exploratory behavior. *Scandinavian Journal of Psychology, 19,* 83-89.

Orleans, P. (1973). Differential cognition of urban residents: Effects of social scale on mapping. In R. M. Downs & D. Stea (Eds.), *Image and environment: Cognitive mapping and spatial behavior* (pp. 115-130). Chicago: Aldine.

Orne, M. (1962). On the social psychology of the psychological experiment. *American Psychologist, 17,* 776-783.

Osgood, C. E. (1971). Explorations in semantic space: A personal diary. *Journal of Social Issues, 27,* 5-64.

Osgood, C. E., Suci, C. J., & Tannenbaum, P. H. (1957). *Measurement of meaning.* Urbana: University of Illinois Press.

Parkes, D., & Thrift, N. (1980). *Times, spaces, and places: A chronogeographic perspective.* New York: John Wiley.

Pearlman, K. T. (1988). Aesthetic regulation and the courts. In J. L. Nasar (Ed.), *Environmental aesthetics: Theory, research, and applications* (pp. 476-492). New York: Cambridge University Press.

Peckham, M. (1976). *Man's rage for chaos: Biology, behavior and the arts.* New York: Schocken.

Peckham, M. (1979). *Explanation and power: The control of human behavior.* Minneapolis: University of Minnesota Press.

Perkins, D. D., Meeks, J. W., & Taylor, R. B. (1992). The physical environment of street blocks and resident perceptions of crime and disorder: Implications for theory and measurement. *Journal of Environmental Psychology, 12,* 21-34.

Perkins, D. D., Wandersman, A., Rich, R. C., & Taylor, R. B. (1993). The physical environment of street crime: Defensible space, territoriality and incivilities. *Journal of Environmental Psychology, 13,* 29-49.

Physical Facilities, Equipment and Library Committee. (1986). *A study of the relationship between the physical environment of the college campus and the quality of academic life.* Columbus: Ohio State University.

Piaget, J. (1962). *Play, dreams and imitation in childhood.* New York: Norton.

Purcell, A. T. (1986). Environmental perception and affect: A schema discrepancy model. *Environment and Behavior, 18,* 3-30.

Purcell, A. T. (1995). Experiencing American and Australian high- and popular-style houses. *Environment and Behavior, 27,* 771-800.

Purcell, A. T., & Nasar, J. L. (1992). Experiencing other people's houses: A model of similarities and differences in environmental experience. *Journal of Environmental Psychology, 12,* 199-211.

Rapoport, A. (1970). Symbolism and environmental design. *International Journal of Symbology, 1,* 1-10.

Rapoport, A. (1977). *Human aspects of urban form.* Oxford: Pergamon.

Rapoport, A. (1990a). *History and precedence in environmental design.* New York: Plenum.

Rapoport, A. (1990b). *The meaning of the built environment: A non-verbal communication approach* (updated ed.). Tucson: University of Arizona Press.

Rapoport, A. (1993). *Cross-cultural studies and urban form.* College Park: University of Maryland.

Rapoport, A., & Hawkes, R. (1970). The perception of urban complexity. *Journal of the American Institute of Planners, 36,* 106-111.

Rosenberg, M. J. (1965). When dissonance fails: On eliminating evaluative apprehension from attitude measurement. *Journal of Personality and Social Psychology, 1,* 28-42.

Rosenthal, R. (1976). *Experimenter effects in behavioral research.* New York: John Wiley.

Ross, L., Amabile, T. M., & Steinmetz, J. L. (1977). Social roles, social control, and biases in social perception processes. *Journal of Personality and Social Psychology, 35,* 485-494.

Royse, D. C. (1969). Social inferences via environmental cues. *Dissertation Abstract International X1969,* 0291. (University Microfilms No. AAI0221503).

Russell, J. A., & Snodgrass, J. (1989). Emotion and environment. In D. Stokols & I. Altman (Eds.), *Handbook of environmental psychology* (Vol. 1, pp. 245-280). New York: John Wiley.

Saarinen, T. F. (1969). *Perception of environment.* (Commission on College Geography Resource Paper No. 5). Washington, DC: Association of American Geographers.

Sadalla, E. K., Verschure, B., & Burroughs, J. (1987). Identity symbolism in housing. *Environment and Behavior, 19,* 569-587.

Sanoff, H. (1977). *Methods of architectural programming.* Stroudsburg, PA: Dowden, Hutchinson and Ross.

Sanoff, H. (1989). Facility programming. In E. H. Zube & G. M. Moore (Eds.), *Advances in environment, behavior, and design* (Vol. 2, pp. 239-286). New York: Plenum.

Sanoff, H., & Sawney, M. (1972). Residential livability: A study of user attitudes toward their residential environment. In W. S. Mitchell (Ed.), *EDRA 3* (Vol. 1, pp. 13-18: 1-10). Los Angeles: University of California Press.

Sauer, L. (1972). The architect and user needs. In W. M. Smith (Ed.), *Behavior, design and policy aspects of human habitats* (pp. 147-170). Green Bay: University of Wisconsin Press.

Seaton, R. W., & Collins, J. B. (1970). Validity and reliability of ratings of simulated buildings. In W. S. Mitchell (Ed.), *Environmental design: Research and practice* (pp. 6-10: 1-12). Los Angeles: University of California Press.

Shafer, E. L., Hamilton, J. F., & Smidt, E. (1969). Natural landscape preferences: A predictive model. *Journal of Leisure Research, 1,* 71-79.

Shafer, E. L., & Richards, T. A. (1974). *A comparison of viewer reactions to outside scenes and photographs of those scenes.* (U.S. Forest Service Research Paper NE 302). Upper Darby, PA: Northeast Forest Experiment Station.

Shirvani, H. (1985). *The urban design process.* New York: Van Nostrand Reinhold.

Shoen, K. (1991). *Districts of pleasantness and danger.* Unpublished manuscript.

Simonton, D. K. (1984). *Genius, creativity, and leadership: Historiometric inquiries.* Cambridge, MA: Harvard University Press.

Simonton, D. K. (1990). *Psychology, science and history: An introduction to historiometry.* New Haven, CT: Yale University Press.

Skogan, W. G., & Maxfield, M. (1981). *Coping with crime: Individual and neighborhood reactions.* Beverly Hills, CA: Sage.

Snodgrass, J., & Russell, J. A. (1986, July). *Mapping the mood of a city.* Paper presented at the 21st Congress of Applied Psychology, Jerusalem, Israel.

Snyder, M., & Swann, W. B. (1978). Hypothesis testing processes in social interaction. *Journal of Personality and Social Psychology, 36,* 1202-1212.

Stamps, A. E. (1990). Use of photographs to simulate environments. A meta-analysis. *Perceptual and Motor Skills, 71,* 907-913.

Stamps, A. E. (1992a). Bootstrap investigation of respondent sample size for environmental preferences. *Perceptual and Motor Skills, 75,* 220-222.

Stamps, A. E. (1992b). Perceptual and preferential effects of photomontage simulations of environments. *Perceptual and Motor Skills, 74,* 675-688.

Stamps, A. E. (1992c). Pre- and post-construction environmental evaluations. *Perceptual and Motor Skills, 75,* 481-482.

Stamps, A. E. (1996). Effect sizes as a lingua franca of environmental aesthetics. In B. Brown & J. L. Nasar (Eds.), *Public/private space* (pp. 151-162). Oklahoma City, OK: Environmental Design Research Association.

Stamps, A. E. (1997, May). *Meta-analysis in environmental research.* Paper presented at the 28th annual conference of the Environmental Design Research Association, Montreal, Canada.

Steinitz, C. (1968). Meaning and the congruence of urban form and activity. *Journal of the American Institute of Planners, 34,* 233-247.

Taylor, R. B. (1989). Towards an environmental psychology of disorder: Delinquency, crime, and fear of crime. In D. Stokols & I. Altman (Eds.), *Handbook of environmental psychology* (Vol. 2, pp. 951-986). New York: John Wiley.

Taylor, R. B., Shumaker, S. A., & Gottfredson, S. D. (1985). Neighborhood-level links between physical features and local sentiments: Deterioration, fear of crime and confidence. *Journal of Architectural and Planning Research, 2,* 261-275.

Thayer, R. L., & Atwood, B. G. (1978). Plant complexity and pleasure in urban and suburban environments. *Environmental Psychology and Nonverbal Behavior, 3,* 67-76.

Thiel, P. (1995). *People, paths and purposes: Notations for a participatory envirotecture.* Seattle: University of Washington Press.

Tunnard, C., & Pushkarev, B. (1981). *Man-made America: Chaos or control.* New York: Harmony.

Tversky, A., & Kahneman, D. (1974). Judgment under uncertainty: Heuristics and biases. *Science, 185,* 1124-1131.

Ulrich, R. S. (1973). Scenery and the shopping trip: The roadside environment as a factor in route choice. *Dissertation Abstracts International, 35*(1) 0346A. (University Microfilms No. AA7415878).

Ulrich, R. S. (1983). Aesthetic and affective response to natural environment. In I. Altman & J. F. Wohlwill (Eds.), *Human behavior and environment: Advances in theory and research* (Vol. 6, pp. 85-125). New York: Plenum.

Ulrich, R. S. (1993). Biophilia and the conservation ethic. In S. R. Kellert & E. O. Wilson (Eds.), *The biophilia hypothesis* (pp. 73-137). Washington, DC: Island.

Ulrich, R. S., Simons, R. F., Losito, B. D., Fiorito, E., Miles, M., & Zelson, M. (1991). Stress recovery during exposure to natural and urban environments. *Journal of Environmental Psychology, 11,* 201-230.

U.S. Department of Commerce, Bureau of Census. (1983). *Census of population, Chapter C general social and economic characteristics, Part 44 Tennessee.* Washington, DC: Government Printing Office.

Verderber, S., & Moore, G. T. (1979). Building imagery: A comparative study of environmental cognition. *Man-Environment Systems, 7,* 332-341.

Vining, J., & Orland, B. (1989). The video advantage: A comparison of two environmental representation techniques. *Journal of Environmental Management, 29,* 275-283.

Ward, L., & Russell, J. A. (1981). The psychological representation of molar environments. *Journal of Experimental Psychology: General, 110,* 121-152.

Warr, M. (1990). Dangerous situations: Social context and fear of victimization. *Social Forces, 68,* 891-907.

Whitfield, T. W. A. (1983). Predicting preference for everyday objects: An experimental confrontation between two theories of aesthetic behavior. *Journal of Environmental Psychology, 3,* 221-237.

Whyte, W. H. (1980). *The social life of small urban spaces.* Washington, DC: Conservation Foundation.

Wilson, J. Q., & Kelling, G. (1982, March). Broken windows. *Atlantic Monthly,* pp. 29-38.

Wilson, M. A., & Canter, D. V. (1990). The development of central concepts during professional education: An example of a multivariate model of the concept of architectural style. *Applied Psychology: An International Review, 39,* 431-455.

Winkel, G., Malek, R., & Thiel, P. (1969). A study of human response to selected roadside environments. In H. Sanoff & S. Cohn (Eds.), *EDRA 1: Proceedings of the 1st Environmental Design Research Association conference* (pp. 224-240). Stroudsburg, PA: Dowden, Hutchinson and Ross.

Wohlwill, J. F. (1974, July). *The place of aesthetics in studies of the environment.* Paper presented at the Symposium on Experimental Aesthetics and Psychology of the Environment at the International Congress of Applied Psychology, Montreal.

Wohlwill, J. F. (1976). Environmental aesthetics: The environment as a source of affect. In I. Altman & J. F. Wohlwill (Eds.), *Human behavior and the environment: Advances in theory and research* (Vol. 1, pp. 37-86). New York: Plenum.

Wohlwill, J. F. (1979). What belongs where: Research on fittingness of man-made structures into natural settings. In T. C. Daniels, E. H. Zube, & B. C. Driver (Eds.), *Assessing amenity resource values* (pp. 48-53). Fort Collins, CO: Rocky Mountain Forest and Range Experimental Station.

Wohlwill, J. F. (1982). The visual impact of development in coastal zone areas. *Coastal Zone Management Journal, 9,* 225-248.

Wohlwill, J. F. (1983). The concept of nature: A psychologist's view. In I. Altman & J. F. Wohlwill (Eds.), *Human behavior and environment: Advances in theory and research* (Vol. 6, pp. 5-37). New York: Plenum.

Wohlwill, J. F., & Harris, G. (1980). Responses to congruity or contrast for man-made features in natural-recreation settings. *Leisure Science, 3,* 349-365.

Wohlwill, J. F., & Kohn, I. (1973). The environment as experienced by the migrant: An adaptation-level view. *Representative Research in Social Psychology, 4,* 135-164.

Zajonc, R. B. (1984). On the primacy of affect. *American Psychologist, 39,* 117-123.

Zube, E. H. (1974). Cross-disciplinary and intermode agreement on the description and evaluation of landscape resources. *Environment and Behavior, 6,* 69-89.

Zube, E. H. (1980). *Environmental evaluation: Perception and public policy.* Monterey, CA: Brooks/Cole.

Zube, E. H., Pitt, D. G., & Anderson, T. W. (1974). *Perception and measurements of scenic resources in the southern Connecticut River valley.* Amherst: University of Massachusetts, Institute for Man and Environment.

Author Index

Subject Index

About the Author

Jack L. Nasar, AICP, is Professor of City and Regional Planning at Ohio State University. A native of New York, he holds degrees in architecture (Washington University), urban planning (master's, New York University), and man–environment relations (PhD, Pennsylvania State University). A Fellow of the American Psychological Association, his other honors and awards include a Lilly Endowment Fellowship, an Ethel Chattel Fellowship, and a College Award for Excellence in Research. His research centers on the relation between the physical environment and human responses and how we can use such information to improve the quality of our cities. He has served as the architectural critic for *The Columbus Dispatch*, has published three books, and has two books forthcoming: *Design by Competition* and *Directions in Person–Environment Research and Practice*. Recent publications include a review of research urban design aesthetics in *Environment and Behavior* (1994), a study of psychological sense of community in the *Journal of the American Planning Association* (1995), an evaluation of design review in the *Journal of Environmental Psychology* (1997), and a study of hot spots of fear in *Environment and Behavior* (1997).